Environmental Engineering

Solved Problems

Second Edition

R. Wane Schneiter, PhD, PE, DEE

Professional Publications, Inc. • Belmont, CA

How to Locate and Report Errata for This Book

At Professional Publications, we do our best to bring you error-free books. But when errors do occur, we want to make sure you can view corrections and report any potential errors you find, so the errors cause as little confusion as possible.

A current list of known errata and other updates for this book is available on the PPI website at **www.ppi2pass.com/errata**. We update the errata page as often as necessary, so check in regularly. You will also find instructions for submitting suspected errata. We are grateful to every reader who takes the time to help us improve the quality of our books by pointing out an error.

ENVIRONMENTAL ENGINEERING SOLVED PROBLEMS
Second Edition

Current printing of this edition: 1

Printing History

edition number	printing number	update
1	2	Minor corrections.
1	3	Minor corrections.
2	1	New title. Minor corrections. Reorganization. Copyright update.

Printed in the United States of America

Professional Publications, Inc.
1250 Fifth Avenue, Belmont, CA 94002
(650) 593-9119
www.ppi2pass.com

Library of Congress Cataloging-in-Publication Data
Schneiter, R. W.
    Environmental engineering solved problems / R. Wane Schneiter.-- 2nd ed.
      p. cm.
    Rev. ed. of: 101 solved environmental engineering problems.
    ISBN-13: 978-1-59126-071-4
    ISBN 10: 1-59126-071-X
    1. Environmental engineering--Problems, exercises, etc. 2. National Council of Examiners for Engineering and Surveying--Examinations--Study guides. I. Schneiter, R.. W. 101 solved environmental engineering problems. II. Title.
TD157.15.S36 2006
628--dc22
                                     2005056492

Table of Contents

Preface

The Principles and Practice of Engineering examination (PE exam) for environmental engineering prepared by the National Council of Examiners for Engineering and Surveying (NCEES) is developed from sample problems submitted by professional engineers. Local licensing boards use the PE exam to uniformly assess the preparation and competency of engineers practicing within their jurisdictions. The uniform application of the PE exam also facilitates reciprocity among many of these jurisdictions.

Because NCEES keeps exam problems confidential, actual NCEES exam problems cannot be included in any study guide, sample PE exam, or other publication, including publications and other materials available from NCEES. However, NCEES does identify the general subject areas covered on the exam. These subject areas form the outline for *Environmental Engineering Solved Problems*.

The problems presented in *Environmental Engineering Solved Problems* cover a broad range of topics relevant to environmental engineering. These problems are both conceptual and practical, and provide varying levels of difficulty in a variety of forms. They are representative of the type and difficulty of problems you will encounter on the PE exam, although the format is slightly different. The PE exam problems are all stand-alone with the solution of each problem being completely independent of others. In this book, some problem solutions are dependent on solutions to preceding problems. This format illustrates the interrelationship of a sequence of questions and should help broaden your understanding of the concepts involved.

Environmental Engineering Solved Problems has been carefully prepared and reviewed to ensure that representative problems are included, that they are appropriate and understandable, and that they have been solved correctly. However, as you work problems, you may discover errors or an alternative, more efficient way to solve a problem. Please bring such discoveries to PPI's attention through **www.ppi2pass.com/errata**. When you do, future editions will include the appropriate corrections and errata sheets will be prepared and published online for those using the current edition. Also, please provide comments about the applicability of specific problems to those you encounter on the exam.

Sometimes you will hear an athlete say, "The will to succeed means nothing without the will to prepare," or "You have nothing to fear if you are prepared." Preparation is the key to success on the PE exam. Use *Environmental Engineering Solved Problems* to prepare for both the content and structure of the exam.

I would like to thank Dr. Brett Borup, PE, of Brigham Young University for his assistance in reviewing the manuscript and providing suggestions for improvement.

R. Wane Schneiter, PhD, PE, DEE
Lexington, VA

Introduction

ORGANIZATION OF THIS BOOK

Although NCEES does not release old PE exams or reveal specific exam problems, it has identified the exam subject areas and their approximate weighting as a percentage of the exam questions. *Environmental Engineering Solved Problems* is organized into four sections that correspond to the four sections of the Environmental PE exam.

1. *Water.* Planning, research, development, project implementation, operations, and monitoring of waste, wastewater, storm water, and natural water systems.

Wastewater	11%
Stormwater	6%
Potable Water	11%
Water Resources	6%

2. *Air.* Planning, research, development, project implementation, operations, and monitoring of air systems.

Ambient Air	8%
Emissions Sources	4%
Control Strategies	8%

3. *Solid, Hazardous, and Special Waste.* Planning, research, development, project implementation, operations, and monitoring of solid and hazardous waste systems.

Municipal Solid Waste (MSW), Commercial, and Industrial Wastes	10%
Hazardous, Special, and Radioactive Wastes	10%

4. *Environmental Assessments, Remediation, and Emergency Response.* Research, development, project implementation, operations, and monitoring of environmental health, safety, and welfare.

Environmental Assessments	8%
Remediation	8%
Public Health and Safety	10%

The problems in this book are organized under 46 scenarios in water, 28 scenarios in air, 28 scenarios in solid, hazardous, and special wastes, and 32 scenarios in environmental assessments, remediation, and emergency response. Many of these problems overlap into more than one of the four main topic areas and include elements of engineering economics in their solutions.

sections	scenarios
I Water	46
II Air	28
III Solid, Hazardous, and Special Waste	28
IV Environmental Assessments, Remediation, and Emergency Response	32
Total scenarios	134

HOW TO USE THIS BOOK

Environmental Engineering Solved Problems should be used to practice solving problems in each of the subject areas covered on the PE exam. The problems are presented in a similar format to those encountered on the exam.

To optimize your study time and obtain the maximum benefit from the practice problems, use the following four-step process.

step 1: Review the problems in each subject area and identify those with which you are least familiar. Work a few of these problems to assess your general understanding of the subject and to identify strengths and weaknesses. When working problems, always make your best attempt to solve the problem before looking at the solutions provided in the book. Then use the solutions to check your work for those problems you are able to solve or to provide guidance in finding solutions to the more difficult problems.

step 2: Focus first on solving problems in those topic areas where the least mastery and greatest weaknesses exist. Begin by locating relevant resource materials, and then work the problems in one subject area at a time. As you work problems, some of these resources will emerge as being more important to you than others. These are the ones you will want to prepare for use when taking the PE exam.

step 3: Use the solved problems as a guide to understanding the general approach to solving problems in each subject area. Although the problems encountered on the PE exam may not be exactly the same as those presented in this book, the approach to solving problems will be the same. It may be useful to make notes in the margins to explain concepts or to reference sources that can help you solve a particular problem. These notes can be consulted while you are taking the exam.

step 4: After you have identified and addressed weaknesses by working problems in those subject areas with which you are least familiar, follow the same procedures outlined above, organizing resource materials and making notations, to work the problems in the remaining subject areas.

Remember that the solution presented for each example problem may represent only one of several methods for obtaining a correct answer. It may also be possible that an alternative method of solving a problem will produce a different, but nonetheless appropriate, answer.

EXAM ORGANIZATION

The PE exam consists of two parts: a morning session and an afternoon session, each lasting four hours. In completing the exam you will need to provide answers to 100 multiple-choice problems, 50 in the morning and 50 in the afternoon. Four possible answers are provided for each problem, but only one of the options is correct. The problems presented in *Environmental Engineering Solved Problems* follow this format. On the PE exam, one point is awarded for each correct answer, but no penalty is assessed for incorrect answers. Therefore, after you have answered problems you are able to solve, you should guess at those you are unable to solve. Don't leave any questions unanswered.

Both the morning and afternoon sessions of the exam are open book. In general, any bound reference material is allowed, including personal notes and sample calculations. Textbooks, handbooks, and other professional reference books are allowed. However, no writing tablets, scratch paper, or other unbound notes or materials are permitted. Battery-operated, silent, nonprinting calculators are allowed. However, each local jurisdiction defines specifically what materials you are permitted to bring with you to the exam. To find out what materials are allowed in the exam in your area, call your state board of registration.

The exam is meant to assess your personal competence without consultation, discussion, or sharing of information with others during the exam period. Do not expect to share any references or to communicate with others while taking the exam.

Information about the state boards can be found at **www.ppi2pass.com/stateboards.html**. More information about the environmental PE exam can be found at **www.ppi2pass.com/enfaqs.html**.

Section I
Water

- Wastewater/Storm Water

- Potable Water

- Water Resources

Wastewater/Storm Water

PROBLEM 1

A treatability study was conducted with wastewater samples using a series of five bench scale bioreactors. The bioreactors were operated without solids recycle. Parameters monitored for each of the reactors are summarized as follows.

test reactor	S_0 (mg/L)	S (mg/L)	θ (d)	X (mg/L)
1	275	8	3.5	137
2	275	15	1.9	128
3	275	21	1.6	135
4	275	32	1.3	130
5	275	44	1.0	124

S_0 is the influent total BOD_5, S is the effluent soluble BOD_5, θ is the hydraulic residence time, and X is the mixed liquor volatile suspended solids.

1.1. What is the value of the yield coefficient?

(A) 0.1 g/g

(B) 0.5 g/g

(C) 1.2 g/g

(D) 1.9 g/g

1.2. What is the value of the endogenous decay rate coefficient?

(A) 0.01 d^{-1}

(B) 0.05 d^{-1}

(C) 0.8 d^{-1}

(D) 1.2 d^{-1}

1.3. What is the value of the growth rate constant?

(A) 0.25 d^{-1}

(B) 4.0 d^{-1}

(C) 25 d^{-1}

(D) 50 d^{-1}

1.4. What is the value of the half-saturation constant?

(A) 0.25 mg/L

(B) 4.0 mg/L

(C) 25 mg/L

(D) 50 mg/L

PROBLEM 2

A complete mix-activated sludge process is being proposed to treat a municipal wastewater at a design flow rate of 8000 m^3/d. Influent soluble BOD must be reduced to an effluent total BOD_5 of 10 mg/L. Influent wastewater characteristics and design requirements include the following.

influent soluble BOD_5 = 235 mg/L = S_0

effluent soluble BOD_5 = 0.3 effluent total BOD_5 = S

X = mixed liquor volatile suspended solids = 3000 mg/L = MLVSS

temperature = 20°C = T

growth rate constant = 2.7 d^{-1}

half-saturation constant = 63 mg/L

endogenous decay rate coefficient = 0.05 d^{-1} = K_d

yield coefficient = 0.23 g/g = Y

design mean cell residence time = 12 d = θ_c

2.1. What is the required hydraulic residence time?

$\theta = ?$

(A) 3.2 h

(B) 8.4 h

(C) 14 h

(D) 56 h

2.2. What is the required bioreactor volume?

(A) 1100 m^3

(B) 2800 m^3

(C) 4700 m^3

(D) 26 000 m^3

2.3. What is the volumetric loading rate?

(A) 0.6 kg/m^3·d

(B) 1.7 kg/m^3·d

(C) 6.0 kg/m^3·d

(D) 17 kg/m^3·d

2.4. What is the food to microorganism ratio?

(A) 0.6 d^{-1}

(B) 1.7 d^{-1}

(C) 6.0 d^{-1}

(D) 17 d^{-1}

2.5. What is the specific substrate utilization rate?

(A) 0.0014 g/g·h

(B) 0.006 g/g·h

(C) 0.01 g/g·h

(D) 0.024 g/g·h

PROBLEM 3

An anaerobic wastewater lagoon has been selected to treat waste at a flow rate of 2300 m³/d with an influent BOD of 670 mg/L. Assume an acceptable winter loading rate of 0.122 kg BOD/m²·d and an acceptable summer loading rate of 0.130 kg BOD/m²·d. The pond depth is set at 4.0 m.

3.1. What is the required surface area for the pond?

(A) 11 800 m²

(B) 12 600 m²

(C) 17 700 m²

(D) 18 900 m²

3.2. What is the pond hydraulic residence time?

(A) 11 d

(B) 22 d

(C) 33 d

(D) 44 d

PROBLEM 4

A municipal wastewater requires nitrification for ammonia removal to 5 mg/L NH_3-N prior to discharge into a nearby stream. The complete mix activated sludge process has been selected for nitrification. The wastewater characteristics and design parameters are

flow rate = 10 000 m³/d

total BOD_5 = 196 mg/L

soluble BOD:total BOD = 0.30

NH_3 = 112 mg/L as N

TKN = 119 mg/L as N

X = MLVSS   mixed liquor volatile suspended solids = 3200 mg/L

pH = 7.4

DO   bioreactor minimum dissolved oxygen = 2.6 mg/L

temperature = 20°C

growth rate constant = 2.4 d⁻¹

K_o half-saturation constant = 21.3 mg/L

endogenous decay rate coefficient = 0.045 d⁻¹

yield coefficient = 0.2 g/g

safety factor = 2.5

4.1. What is the design mean cell residence time for nitrification?

(A) 8 h

(B) 24 h

(C) 60 h

(D) 130 h

4.2. What is the hydraulic residence time for nitrification?

(A) 0.75 h

(B) 1.5 h

(C) 4.0 h

(D) 7 h

4.3. What is the bioreactor volume for nitrification?

(A) 320 m³

(B) 620 m³

(C) 1700 m³

(D) 2900 m³

PROBLEM 5

An existing industrial wastewater treatment plant receives pulp and paper wastewater at 15 000 m³/d with a BOD_5 of 240 mg/L. The plant produces about 1400 kg/d of dry sludge as volatile solids and is able to satisfy effluent criteria for BOD_5 of 30 mg/L. The current aeration equipment is approaching the end of its design life and is scheduled for replacement with low-speed surface aerators. Conditions affecting aeration requirements include the following.

BOD_u = 1.38 BOD_5

aerator oxygen transfer rate = 1.8 kg/kW·h at 20°C and 0 mg/L dissolved oxygen

elevation = 1000 m above mean sea level

salinity = 20 000 mg/L

temperature = 22°C

5.1. What is the daily mass of dissolved oxygen required?

(A) 1200 kg O_2/d

(B) 2400 kg O_2/d

(C) 3200 kg O_2/d

(D) 4400 kg O_2/d

5.2. What is the dissolved oxygen transfer rate under field conditions?

(A) 0.68 kg/kW·h
(B) 0.92 kg/kW·h
(C) 1.4 kg/kW·h
(D) 2.0 kg/kW·h

5.3. What is the power required at the aerator?

(A) 71 kW
(B) 150 kW
(C) 200 kW
(D) 270 kW

PROBLEM 6

An activated sludge process was designed and operates with the following characteristics.

flow rate = 5000 m^3/d
influent BOD$_5$ = 204 mg/L
effluent BOD$_5$ = 20 mg/L
soluble BOD:total BOD = 0.37
influent TSS = 40 mg/L
effluent TSS = 20 mg/L
mixed liquor suspended solids = 3100 mg/L
wasted suspended solids = 15 000 mg/L
VSS:TSS = 0.80
yield coefficient = 0.53 g/g
endogenous decay rate coefficient = 0.048 d^{-1}
bioreactor hydraulic residence time = 6 h
design mean cell residence time = 10 d

6.1. What is the daily mass of biomass produced in the bioreactor?

(A) 330 kg/d
(B) 350 kg/d
(C) 490 kg/d
(D) 520 kg/d

6.2. What is the daily volume of sludge wasted?

(A) 20 m^3/d
(B) 80 m^3/d
(C) 200 m^3/d
(D) 600 m^3/d

6.3. What is the daily mass of sludge wasted?

(A) 300 kg/d
(B) 1200 kg/d
(C) 3000 kg/d
(D) 9000 kg/d

6.4. What is the recirculated solids flow rate?

(A) 130 m^3/d
(B) 850 m^3/d
(C) 1300 m^3/d
(D) 1700 m^3/d

6.5. What is the daily mass of solids recirculated?

(A) 1900 kg/d
(B) 13 000 kg/d
(C) 20 000 kg/d
(D) 26 000 kg/d

PROBLEM 7

A food processing plant produces a high-strength wastewater that is to be pretreated using a super high-rate, plastic media trickling filter prior to discharge to a municipal sewer. The plant produces a continuous flow of 1600 m^3/d of wastewater at a temperature of 16°C. The wastewater BOD of 434 mg/L must be reduced to 50 mg/L to meet discharge requirements. The specified parameters for the trickling filter are

treatability constant, k_{20} = 0.075(gpm)^n/ft^2 with $n = 0.5$
temperature correction coefficient, $\theta = 1.06$
filter depth, $D = 3.7$ m
media coefficient, $x = 0.3$
recirculation factor, $\alpha_R = 2$
dosing rate, DR = 20 cm/pass of the recirculation arm

7.1. What cross-sectional surface area is required for the filter?

(A) 64 m^2
(B) 190 m^2
(C) 370 m^2
(D) 480 m^2

7.2. What is the hydraulic loading rate?

(A) 0.027 m^3/m^2·min
(B) 0.052 m^3/m^2·min
(C) 0.12 m^3/m^2·min
(D) 0.23 m^3/m^2·min

7.3. What is the organic loading rate?

(A) 1.2 kg/m^3·d

(B) 1.6 kg/m^3·d

(C) 2.9 kg/m^3·d

(D) 10 kg/m^3·d

7.4. What is the distribution arm rotational speed if a two-arm distributor is used?

(A) 0.08 rpm

(B) 0.13 rpm

(C) 1.7 rpm

(D) 2.6 rpm

PROBLEM 8

A secondary clarifier accepts effluent from a bioreactor at a flow rate of 8300 m^3/d and total suspended solids of 1600 mg/L. The solids flux for the suspension is 2.6 kg/m^2·h, and the particle settling velocity is 1.29 m/h.

8.1. What is the solids loading rate?

(A) 1300 kg/d

(B) 1600 kg/d

(C) 13 000 kg/d

(D) 17 000 kg/d

8.2. What is the required surface area based on solids flux?

(A) 130 m^2

(B) 210 m^2

(C) 270 m^2

(D) 430 m^2

8.3. What is the required surface area based on particle settling velocity?

(A) 130 m^2

(B) 210 m^2

(C) 270 m^2

(D) 430 m^2

8.4. What is the design surface area?

(A) 130 m^2

(B) 210 m^2

(C) 270 m^2

(D) 430 m^2

8.5. What is the design overflow rate?

(A) 0.8 m^3/m^2·h

(B) 1.3 m^3/m^2·h

(C) 1.6 m^3/m^2·h

(D) 2.6 m^3/m^2·h

PROBLEM 9

A thickener is required for a wastewater suspension that follows type III settling characteristics as represented by the settling column test results presented in the following table. The mixed suspension TSS is 5860 mg/L.

settling time (min)	solid-liquid interface height (cm)
0	100
5	93
10	85
20	74
30	68
40	62
50	59
60	57
70	53
80	52
90	50

The thickener will receive 500 m^3/d of flow and should be capable of providing a thickened solids concentration of 11 000 mg/L.

9.1. What is the surface area required for thickening?

(A) 7.5 m^2

(B) 13 m^2

(C) 22 m^2

(D) 75 m^2

9.2. What is the particle settling velocity?

(A) 0.01 m/h

(B) 0.1 m/h

(C) 1.0 m/h

(D) 10 m/h

9.3. What is the surface area required for clarification?

(A) 7.5 m^2

(B) 9.3 m^2

(C) 13 m^2

(D) 75 m^2

9.4. What is the design surface area?

(A) 7.5 m²
(B) 13 m²
(C) 22 m²
(D) 75 m²

9.5. What is the design overflow rate?

(A) 4 m³/m²·d
(B) 7 m³/m²·d
(C) 23 m³/m²·d
(D) 40 m³/m²·d

9.6. What is the solids volume wasted through the thickener?

(A) 32 m³/d
(B) 270 m³/d
(C) 500 m³/d
(D) 2900 m³/d

10.2. What is the daily mass of ozone used to treat the wastewater?

(A) 520 kg/d
(B) 5200 kg/d
(C) 52 000 kg/d
(D) 520 000 kg/d

10.3. What is the annual cost for chlorine?

(A) $50,000/yr
(B) $170,000/yr
(C) $450,000/yr
(D) $16,000,000/yr

10.4. What is the annual cost for ozone?

(A) $110,000/yr
(B) $310,000/yr
(C) $11,000,000/yr
(D) $320,000,000/yr

PROBLEM 10

A precious metals mine produces 20 000 m³/d of wastewater containing free cyanide (CN^-) at 1400 mg/L. The wastewater can meet discharge criteria if the free cyanide is oxidized to cyanate (CNO^-). The two choices available for use as an oxidizing reagent are chlorine gas (Cl_2) and ozone (O_3). The ozone generator produces 3.5% O_3 with a power requirement of 14 kW·h/kg of ozone produced. Electrical power costs $0.042/kW·h. Chlorine gas, at a purity of 99.8%, is available at $584/1000 kg, which includes transportation to the site. The oxidation treatment process will operate continuously, 24 h daily and 365 d annually.

The reaction equation for cyanide oxidation using ozone is

$$CN^- + O_3 \rightarrow CNO^- + O_2$$

The reaction equation for cyanide oxidation using chlorine gas is

$$CN^- + Cl_2 + H_2O \longrightarrow CNO^- + 2H^+ + 2Cl^-$$

10.1. What is the daily mass of chlorine gas used to treat the wastewater?

(A) 770 kg/d
(B) 7700 kg/d
(C) 77 000 kg/d
(D) 770 000 kg/d

PROBLEM 11

Air stripping has been selected to remove volatile organic chemicals from a water supply. The water target chemical characteristics and stripping tower design constraints are as follows.

target chemical
  influent concentration = 1.57 mg/L
  required effluent concentration = 0.05 mg/L
  vapor pressure = 0.11 atm at 20°C
  solubility in water = 1250 mg/L at 20°C
  molecular weight = 87 g/mol
  mass transfer coefficient = 0.019 s⁻¹

design constraints
  flow rate = 3000 m³/d
  temperature = 20°C = 293K
  stripping factor = 4
  maximum column diameter = 1.0 m

11.1. What is the value of Henry's constant in unitless form?

(A) 8.8×10^{-5}
(B) 7.7×10^{-3}
(C) 0.32
(D) 93

11.2. What is the air:water ratio?

(A) 2.5

(B) 13

(C) 25

(D) 50

11.3. What is the transfer unit height?

(A) 1.5 m

(B) 2.5 m

(C) 3.5 m

(D) 4.5 m

11.4. How many transfer units are required?

(A) 1.5

(B) 2.5

(C) 3.5

(D) 4.5

11.5. What is the packing height?

(A) 6.75 m

(B) 8.75 m

(C) 11.25 m

(D) 12.25 m

PROBLEM 12

Monitoring at a wastewater treatment plant reveals the daily flow variations summarized in the following table.

period	average flow (m^3/period)
0000–0200	275
0200–0400	389
0400–0600	621
0600–0800	1340
0800–1000	1383
1000–1200	1312
1200–1400	1098
1400–1600	1027
1600–1800	1084
1800–2000	886
2000–2200	259
2200–2400	326

12.1. What storage capacity is required to equalize the flow?

(A) 2500 m^3

(B) 5500 m^3

(C) 11 000 m^3

(D) 20 000 m^3

12.2. When will the storage tank be at its peak level?

(A) 0500 h

(B) 1000 h

(C) 1900 h

(D) 2400 h

12.3. When will the storage tank be at its minimum level?

(A) 0500 h

(B) 1000 h

(C) 1900 h

(D) 2400 h

PROBLEM 13

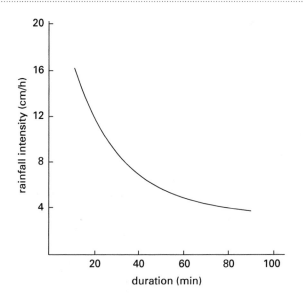

A watershed occupies a 30 ha site. 18 ha of the site have been cleared and are used for pasture land; 1 ha is occupied by farm buildings, a house, and paved surfaces; the remaining 11 ha are woodland. The average land slope is 2.1%. Because the site is upland from a residential development, the rainfall runoff from the site is collected in a catchment that discharges directly to a culvert. The overland flow distance to the catchment is 212 m.

The 20 yr storm is characterized by the intensity duration curve presented in the figure.

13.1. What is the weighted average runoff coefficient for the watershed?

(A) 0.07

(B) 0.18

(C) 0.34

(D) 0.62

13.2. What is the total time of concentration for the watershed?

(A) 19 min

(B) 34 min

(C) 87 min

(D) 160 min

13.3. What is peak flow rate to the culvert entrance for the 20 yr storm?

(A) 0.44 m^3/s

(B) 1.1 m^3/s

(C) 2.1 m^3/s

(D) 3.9 m^3/s

PROBLEM 14

Local public works personnel have constructed a rough earthen ditch to divert frequent floodwater away from residential property. The ditch has proven to be effective and the residents want the city to line the ditch with asphalt. When in use, the water depth in the ditch averages 0.8 m. The ditch is 800 m long on a slope of 1.8% with 1:1 side slopes and a 0.5 m wide bottom.

14.1. What is the flow velocity in the earthen ditch?

(A) 1.0 m/s

(B) 3.0 m/s

(C) 10 m/s

(D) 20 m/s

14.2. What is the flow discharge in the earthen ditch?

(A) 1.0 m^3/s

(B) 3.0 m^3/s

(C) 10 m^3/s

(D) 20 m^3/s

14.3. If the dimensions and configuration of the ditch remain unchanged, what will be the average depth of water in the ditch after it is lined with asphalt?

(A) 0.5 m

(B) 0.6 m

(C) 0.9 m

(D) 1.1 m

PROBLEM 15

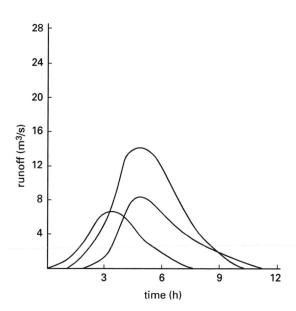

The figure presents 10 min unit hydrographs developed for a 30 min storm.

15.1. What is the peak discharge for the storm?

(A) 7 m^3/s

(B) 9 m^3/s

(C) 15 m^3/s

(D) 26 m^3/s

15.2. How long after the beginning of the storm does peak discharge occur?

(A) 3 h

(B) 5 h

(C) 7 h

(D) 9 h

15.3. What is the total flow from the storm runoff?

(A) 7000 m^3

(B) 18 000 m^3

(C) 170 000 m^3

(D) 420 000 m^3

PROBLEM 16

20 yr of 24 h annual peak discharge values are summarized as follows.

year	annual peak discharge (m^3/s)
1975	112
1976	94
1977	54
1978	49
1979	42
1980	51
1981	128
1982	103
1983	88
1984	96
1985	65
1986	79
1987	83
1988	71
1989	57
1990	89
1991	62
1992	53
1993	64
1994	92

16.1. What is the peak discharge expected for the average 2 yr, 24 h storm?

   (A) 42 m^3/s

   (B) 46 m^3/s

   (C) 50 m^3/s

   (D) 54 m^3/s

16.2. What is the peak discharge expected for the average 10 yr, 24 h storm?

   (A) 65 m^3/s

   (B) 75 m^3/s

   (C) 81 m^3/s

   (D) 90 m^3/s

PROBLEM 17

Leaks have been discovered in piping from a fuel storage depot. The leaking pipes convey no. 2 fuel oil to tanker truck fill stations at the facility. The site is characterized by an unconfined aquifer having an average hydraulic conductivity for the aquifer of 0.42 m/d, a gradient of 0.022, and an effective porosity of 0.28. The average groundwater temperature is 8°C. The density and dynamic viscosity of no. 2 fuel are 900 kg/m^3 and 6.5×10^{-3} kg/m·s, respectively.

17.1. What is the intrinsic permeability of the soil?

   (A) 0.42 m/d

   (B) 0.060 m^2

   (C) 5.0×10^{-7} s

   (D) 6.9×10^{-13} m^2

17.2. What is the hydraulic conductivity of the aquifer for no. 2 fuel oil as a nonaqueous phase liquid?

   (A) 9.4×10^{-7} m/d

   (B) 8.4×10^{-4} m/d

   (C) 0.081 m/d

   (D) 0.42 m/d

17.3. What is the average actual velocity of the no. 2 fuel oil as a nonaqueous phase liquid?

   (A) 3.3×10^{-2} m/d

   (B) 6.4×10^{-3} m/d

   (C) 6.6×10^{-5} m/d

   (D) 7.4×10^{-8} m/d

PROBLEM 18

Bench scale aeration tests were performed on a sample of water that is being evaluated for air stripping to remove VOCs. The target VOC is present in the water sample before aeration at 990 μg/L. The tests have produced the following data.

elapsed time of aeration (s)	target VOC effluent concentration (μg/L)
30	497
60	251
90	124
120	63
180	14
240	3

18.1. What is the reaction order?

   (A) zero

   (B) first

   (C) second

   (D) pseudo first

18.2. What is the value of the mass transfer coefficient?

   (A) 0.023 s^{-1}

   (B) 0.092 s^{-1}

   (C) 0.21 s^{-1}

   (D) 0.82 s^{-1}

18.3. How long will be required to reduce the target VOC concentration to 1.0 μg/L?

(A) 260 s

(B) 300 s

(C) 390 s

(D) 470 s

PROBLEM 19

Wastewater samples were prepared and incubated at 20°C for 5 d for BOD analysis. Sample dilutions and initial and final dissolved oxygen concentrations are summarized in the following table.

bottle	sample volume (mL)	initial dissolved oxygen (mg/L)	final dissolved oxygen (mg/L)
1	5	9.3	7.7
2	10	9.2	6.6
3	15	9.1	5.2
4	20	9.1	4.1
5	30	8.9	1.8

The temperature correction coefficient is 1.047, and the reaction rate coefficient at 20°C is 0.40 d^{-1}.

19.1. What is the BOD_5 at 20°C?

(A) 71 mg/L

(B) 77 mg/L

(C) 80 mg/L

(D) 96 mg/L

19.2. What is the BOD_u?

(A) 82 mg/L

(B) 89 mg/L

(C) 93 mg/L

(D) 110 mg/L

19.3. What is the BOD_7 at 15°C?

(A) 73 mg/L

(B) 80 mg/L

(C) 85 mg/L

(D) 98 mg/L

Wastewater/Storm Water Solutions

SOLUTION 1

1.1. The yield coefficient and the endogenous decay rate coefficient are evaluated by plotting the equation or through linear regression. The following table presents the data used to perform the linear regression or to plot the equation presented in the accompanying figure.

θ = hydraulic residence time, d
θ_c = mean cell residence time, d
   ($\theta_c = \theta$ when reactors are operated without solids recycle)
Y = yield coefficient, g/g
S_0, S = influent and effluent BOD, mg/L
k_d = endogenous decay rate coefficient, d^{-1}
X = mixed liquor volatile suspended solids, mg/L

$$\frac{1}{\theta_c} = \frac{Y(S_0 - S)}{\theta X} - k_d$$

test reactor	$1/\theta_c$ (d^{-1})	$S_0 - S$ (mg/L)	θX (mg·d/L)	$(S_0 - S)/\theta X$ (d^{-1})
1	0.29	267	480	0.56
2	0.53	260	243	1.1
3	0.63	254	216	1.2
4	0.77	243	169	1.4
5	1.0	231	124	1.9

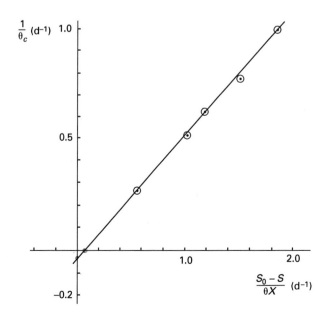

From the figure (or through linear regression), Y is the slope.

$$Y = \frac{1.0\ d^{-1} - 0.29\ d^{-1}}{1.9\ d^{-1} - 0.56\ d^{-1}} = 0.53 \quad (0.5)$$

By definition, units for Y are, g/g.

$$Y = \boxed{0.5\ \text{g/g}}$$

The answer is B.

1.2. From the figure (or through linear regression), k_d is the y-intercept.

$$Y = \frac{1.0\ d^{-1} - k_d}{1.9\ d^{-1} - 0} = 0.5$$

$$k_d = 0.1\ d^{-1} - (0.5)(1.9\ d^{-1})$$

$$= \boxed{0.05\ d^{-1}}$$

The answer is B.

1.3. The growth rate constant and the half-saturation constant are evaluated by plotting the equation or through linear regression. The following table presents the data used to perform the linear regression or to plot the equation presented in the accompanying figure.

k = growth rate constant, d^{-1}
K_s = half-saturation constant, mg/L

$$\frac{\theta X}{S_0 - S} = \frac{K_s}{kS} + \frac{1}{k}$$

test reactor	$1/S$ (L/mg)	$S_0 - S$ (mg/L)	θX (mg/L·d)	$\theta X/(S_0 - S)$ (d)
1	0.125	267	480	1.80
2	0.067	260	243	0.93
3	0.048	254	216	0.85
4	0.031	243	169	0.70
5	0.023	231	124	0.54

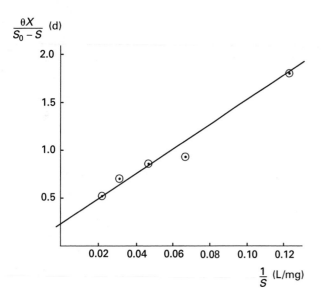

From the figure (or through linear regression), $1/k$ is the y-intercept.

$$\frac{1}{k} = 0.25 \text{ d}$$

$$k = \boxed{4.0 \text{ d}^{-1}}$$

The answer is B.

1.4. From the figure (or through linear regression), K_s/k is the slope.

$$\frac{K_s}{k} = \frac{1.80 \text{ d} - 0.54 \text{ d}}{0.125 \dfrac{\text{L}}{\text{mg}} - 0.023 \dfrac{\text{L}}{\text{mg}}}$$

$$= 12.4 \text{ mg·d/L}$$

$$K_s = \left(12.4 \frac{\text{mg·d}}{\text{L}} \right) (4.0 \text{ d}^{-1})$$

$$= \boxed{49.6 \text{ mg/L} \quad (50 \text{ mg/L})}$$

The answer is D.

SOLUTION 2

2.1. For complete mix-activated sludge,

θ = hydraulic residence time, h
θ_c = mean cell residence time = 12 d
Y = yield coefficient = 0.23 g/g
S_0 = influent soluble BOD_5 = 235 mg/L
S = effluent soluble BOD_5, mg/L
X = mixed liquor volatile suspended solids
     = 3000 mg/L

k_d = endogenous decay rate coefficient = 0.05 d^{-1}
$S = (0.3)(10 \text{ mg/L}) = 3 \text{ mg/L}$

$$\theta = \frac{\theta_c Y (S_0 - S)}{X (1 + k_d \theta_c)}$$

$$= \frac{(12 \text{ d}) \left(0.23 \dfrac{\text{g}}{\text{g}} \right) \left(235 \dfrac{\text{mg}}{\text{L}} - 3 \dfrac{\text{mg}}{\text{L}} \right) \left(24 \dfrac{\text{h}}{\text{d}} \right)}{\left(3000 \dfrac{\text{mg}}{\text{L}} \right) \left(1 + (0.05 \text{ d}^{-1})(12 \text{ d}) \right)}$$

$$= \boxed{3.2 \text{ h}}$$

The answer is A.

2.2. V = bioreactor volume, m^3
Q = influent flow rate = 8000 m^3/d

$$V = Q\theta$$

$$= \left(8000 \frac{\text{m}^3}{\text{d}} \right) (3.2 \text{ h}) \left(\frac{1 \text{ d}}{24 \text{ h}} \right)$$

$$= \boxed{1067 \text{ m}^3 \quad (1100 \text{ m}^3)}$$

The answer is A.

2.3. VLR = volumetric loading rate, kg/m^3·d

$$\text{VLR} = \frac{S_0 Q}{V}$$

$$= \frac{\left(235 \dfrac{\text{mg}}{\text{L}} \right) \left(8000 \dfrac{\text{m}^3}{\text{d}} \right) \times \left(10^{-6} \dfrac{\text{kg}}{\text{mg}} \right) \left(1000 \dfrac{\text{L}}{\text{m}^3} \right)}{(1100 \text{ m}^3)}$$

$$= \boxed{1.7 \text{ kg/m}^3 \text{·d}}$$

The answer is B.

2.4. F/M = food to microorganism ratio, d^{-1}

$$\frac{F}{M} = \frac{S_0}{\theta X}$$

$$= \frac{\left(235 \dfrac{\text{mg}}{\text{L}} \right) \left(24 \dfrac{\text{h}}{\text{d}} \right)}{(3.2 \text{ h}) \left(3000 \dfrac{\text{mg}}{\text{L}} \right)}$$

$$= \boxed{0.59 \text{ d}^{-1} \quad (0.6 \text{ d}^{-1})}$$

The answer is A.

2.5. U = specific substrate utilization rate, mg/L·h

$$U = \frac{S_0 - S}{\theta X}$$

$$= \frac{235 \frac{mg}{L} - 3 \frac{mg}{L}}{(3.2 \text{ h}) \left(3000 \frac{mg}{L}\right)}$$

$$= 0.024 \text{ h}^{-1}$$

By definition, units for U are, g/g·h.

$$U = \boxed{0.024 \text{ g/g·h}}$$

The answer is D.

SOLUTION 3

3.1. The winter loading rate controls design.

$$\text{surface area} = \frac{\left(2300 \frac{m^3}{d}\right) \left(670 \frac{mg}{L}\right)}{\times \left(1000 \frac{L}{m^3}\right) \left(10^{-6} \frac{kg}{mg}\right)}{0.122 \frac{kg}{m^2 \cdot d}}$$

$$= \boxed{12\,631 \text{ m}^2 \quad (12\,600 \text{ m}^2)}$$

The answer is B.

3.2. The hydraulic residence time is

$$t = \frac{(12\,600 \text{ m}^2)(4.0 \text{ m})}{2300 \frac{m^3}{d}} = \boxed{21.9 \text{ d} \quad (22 \text{ d})}$$

The answer is B.

SOLUTION 4

4.1. μ_m = maximum growth rate, d^{-1}
Y = yield coefficient = 0.2 g/g
k = growth rate constant = 2.4 d^{-1}

$$\mu_m = Yk$$

$$= \left(0.2 \frac{g}{g}\right)(2.4 \text{ d}^{-1})$$

$$= 0.48 \text{ d}^{-1}$$

μ_m' = corrected maximum growth rate, d^{-1}
T = temperature = 20°C

DO = minimum dissolved oxygen concentration
   in bioreactor
   = 2.6 mg/L
K_o = half-saturation constant for dissolved oxygen,
   typical value
   = 1.3 mg/L

$$\mu_m' = \frac{\mu_m \left(e^{(0.098)(T-15)}\right)(DO)\left(1 - (0.833)(7.2 - pH)\right)}{K_o + DO}$$

$$= \frac{\begin{array}{c}(0.48 \text{ d}^{-1}) \left(e^{(0.098)(20-15)}\right) \left(2.6 \frac{mg}{L}\right) \\ \times \left(1 - (0.833)(7.2 - 7.4)\right)\end{array}}{1.3 \frac{mg}{L} + 2.6 \frac{mg}{L}}$$

$$= 0.61 \text{ d}^{-1}$$

θ_c^d = design mean cell residence time for
   nitrification, h
θ_c^m = minimum mean cell residence time, d
N_o = influent ammonia concentration
   = 112 mg/L as N
k_d = endogenous decay rate coefficient = 0.045 d^{-1}
K_s = half-saturation constant = 21.3 mg/L
SF = safety factor = 2.5

$$\frac{1}{\theta_c^m} = \frac{\mu_m' N_o}{K_s + N_o} - k_d$$

$$= \frac{(0.61 \text{ d}^{-1}) \left(112 \frac{mg}{L}\right)}{21.3 \frac{mg}{L} + 112 \frac{mg}{L}} - 0.045 \text{ d}^{-1}$$

$$= 0.47 \text{ d}^{-1}$$

$$\theta_c^m = 2.1 \text{ d}$$

$$\theta_c^d = (SF)\theta_c^m = (2.5)(2.1 \text{ d}) \left(24 \frac{h}{d}\right)$$

$$= \boxed{126 \text{ h} \quad (130 \text{ h})}$$

The answer is D.

4.2. θ = hydraulic residence time, h
N = effluent ammonia concentration
   = 5 mg/L as N
X_N = nitrifying biomass concentration, mg/L

$$\theta = \frac{Y(N_o - N)}{X_N \left(\frac{1}{\theta_c^d} + k_d\right)}$$

When BOD:TKN = 1.6, the nitrifying fraction of the biomass is approximately 0.17.

$$X_N = (0.17)\left(3200 \ \frac{mg}{L}\right) = 544 \ mg/L$$

$$\theta_c^d = (130 \ h)\left(\frac{1 \ d}{24 \ h}\right) = 5.42 \ d$$

$$\theta = \frac{\left(0.2 \ \frac{g}{g}\right)\left(112 \ \frac{mg}{L} - 5 \ \frac{mg}{L}\right)\left(24 \ \frac{h}{d}\right)}{\left(544 \ \frac{mg}{L}\right)\left(\frac{1}{5.42 \ d} + 0.045 \ d^{-1}\right)}$$

$$= \boxed{4.1 \ h \quad (4.0h)}$$

The answer is C.

4.3. The bioreactor volume is

$$(4.0 \ h)\left(10\,000 \ \frac{m^3}{d}\right)\left(\frac{1 \ d}{24 \ h}\right) = \boxed{1667 \ m^3 \quad (1700 \ m^3)}$$

The answer is C.

SOLUTION 5

5.1. Q = water flow rate = 15 000 m³/d
S_0 = influent BOD₅ = 240 mg/L
S = effluent BOD₅ = 30 mg/L
f = ratio of BOD$_u$:BOD₅ = 1.38
X_p = volatile solids production rate
  = 1400 kg/d
$\dot{m}$ = oxygen required, kg/d

$$\dot{m} = Q(S_0 - S)f - 1.42X_p$$

The daily O₂ requirement is

$$\left(15\,000 \ \frac{m^3}{d}\right)\left(240 \ \frac{mg}{L} - 30 \ \frac{mg}{L}\right)$$

$$\times \left(1000 \ \frac{L}{m^3}\right)\left(10^{-6} \ \frac{kg}{mg}\right)(1.38)$$

$$- (1.42)\left(1400 \ \frac{kg}{d}\right)$$

$$= \boxed{2359 \ kg/d \quad (2400 \ kg/d)}$$

The answer is B.

5.2. N = oxygen transfer rate under field conditions, kg/kW·h
N_o = oxygen transfer rate in water at 20°C and 0 mg/L dissolved oxygen
  = 1.8 kg/kW·h
β = salinity-surface tension correction factor, unitless
  = 1.0 (commonly selected value)
$C_{w\text{-}alt}$ = tap water oxygen saturation concentration at a given temperature and altitude

$$N = N_o\left(\frac{\beta C_{w\text{-}alt} - C_L}{C_{s20}}\right)\alpha\left(1.024^{(T-20)}\right)$$

$$C_{w\text{-}alt} = \left(7.77 \ \frac{mg}{L}\right)(0.88) = 6.8 \ mg/L$$

The dissolved oxygen concentration in water at 22°C and salinity of 20 000 mg/L is 7.77 mg/L and the elevation correction factor is 0.88.

C_L = operating oxygen concentration
  = 2.0 mg/L (minimum required for aerobic treatment)
α = wastewater oxygen-transfer correction factor, unitless
  = 0.68 (typical for pulp and paper)
T = wastewater operating temperature = 22°C
C_{s20} = tap water oxygen saturation concentration at 20°C and salinity of 0 mg/L = 9.08 mg/L

$$N = \frac{\left(1.8 \ \frac{kg}{kW\cdot h}\right)\left((1)\left(6.8 \ \frac{mg}{L}\right) - 2.0 \ \frac{mg}{L}\right)}{9.08 \ \frac{mg}{L}} \times (0.68)\left(1.024^{(22-20)}\right)$$

$$= \boxed{0.678 \ kg/kW\cdot h \quad (0.68 \ kg/kW\cdot h)}$$

The answer is A.

5.3. The power required at the aerator is

$$P = \frac{\dot{m}}{N} = \frac{\left(2400 \ \frac{kg}{d}\right)\left(\frac{1 \ d}{24 \ h}\right)}{0.68 \ \frac{kg}{kW\cdot h}}$$

$$= \boxed{147 \ kW \quad (150 \ kW)}$$

The answer is B.

SOLUTION 6

6.1. Y_{obs} = observed yield coefficient, g/g
Y = yield coefficient = 0.53 g/g
k_d = endogenous decay rate coefficient
  = 0.048 d^{-1}
θ_c = design mean cell residence time = 10 d

$$Y_{obs} = \frac{Y}{1 + k_d \theta_c} = \frac{0.53 \frac{g}{g}}{1 + (0.048 \text{ d}^{-1})(10 \text{ d})} = 0.36 \text{ g/g}$$

X_p = biomass production rate, kg/d
S_0 = influent total BOD$_5$ = 204 mg/L
S = effluent soluble BOD$_5$ = (0.37)(20 mg/L)
  = 7.4 mg/L
Q = wastewater flow rate = 5000 m^3/d

$$X_p = Y_{obs}(S_0 - S)Q$$

$$= \left(0.36 \frac{g}{g}\right)\left(204 \frac{mg}{L} - 7.4 \frac{mg}{L}\right)\left(5000 \frac{m^3}{d}\right)$$

$$\times \left(1000 \frac{L}{m^3}\right)\left(10^{-6} \frac{kg}{mg}\right)$$

$$= \boxed{354 \text{ kg/d} \quad (350 \text{ kg/d})}$$

The answer is B.

6.2. V = bioreactor volume, m^3
X = mixed liquor suspended solids
  = 3100 mg/L
Q_w = wasted solids flow rate, m^3/d
X_u = wasted suspended solids = 15 000 mg/L
X_e = clarifier effluent TSS = 20 mg/L

$$V = \left(5000 \frac{m^3}{d}\right)(6 \text{ h})\left(\frac{1 \text{ d}}{24 \text{ h}}\right) = 1250 \text{ m}^3$$

$$\theta_c = \frac{VX}{Q_w X_u + (Q - Q_w)(X_e)}$$

$$Q_w = \frac{\dfrac{VX}{\theta_c} - QX_e}{X_u - X_e}$$

$$= \frac{\dfrac{(1250 \text{ m}^3)\left(3100 \frac{mg}{L}\right)}{10 \text{ d}} - \left(5000 \frac{m^3}{d}\right)\left(20 \frac{mg}{L}\right)}{15\,000 \frac{mg}{L} - 20 \frac{mg}{L}}$$

$$= \frac{287\,500 \frac{m^3 \cdot mg}{L \cdot d}}{15\,000 \frac{mg}{L} - 20 \frac{mg}{L}}$$

$$= \boxed{19.2 \text{ m}^3/\text{d} \quad (20 \text{ m}^3/\text{d})}$$

The answer is A.

6.3. The daily sludge mass wasted is

$$\left(15\,000 \frac{mg}{L}\right)\left(20 \frac{m^3}{d}\right)$$

$$\times \left(10^{-6} \frac{kg}{mg}\right)\left(1000 \frac{L}{m^3}\right) = \boxed{300 \text{ kg/d}}$$

The answer is A.

6.4. To determine the recirculated solids flow rate, Q_R, perform a mass balance around the bioreactor as depicted in the figure.

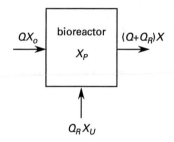

X_o = influent TSS = 40 mg/L
X_p = biomass production rate based on total solids

$$= \frac{350 \frac{kg}{d}}{0.80}$$

$$= 438 \text{ kg/d}$$

$$\text{inputs} = \text{outputs}$$

$$QX_o + X_p + Q_R X_u = QX + Q_R X$$

$$Q_R = \frac{Q(X_o - X) + X_p}{(X - X_u)}$$

$$= \frac{\begin{array}{c}\left(5000 \frac{m^3}{d}\right)\left(40 \frac{mg}{L} - 3100 \frac{mg}{L}\right)\left(10^{-6} \frac{kg}{mg}\right) \\ \times \left(1000 \frac{L}{m^3}\right) + 438 \frac{kg}{d}\end{array}}{\begin{array}{c}\left(3100 \frac{mg}{L} - 15\,000 \frac{mg}{L}\right) \\ \times \left(10^{-6} \frac{kg}{mg}\right)\left(1000 \frac{L}{m^3}\right)\end{array}}$$

$$= \boxed{1249 \text{ m}^3/\text{d} \quad (1300 \text{ m}^3/\text{d})}$$

The answer is C.

6.5. The recirculated solids daily mass is

$$\left(1300 \frac{m^3}{d}\right)\left(15\,000 \frac{mg}{L}\right)\left(10^{-6} \frac{kg}{mg}\right)\left(1000 \frac{L}{m^3}\right)$$

$$= \boxed{19\,500 \text{ kg/d} \quad (20\,000 \text{ kg/d})}$$

The answer is C.

SOLUTION 7

7.1. The equations are empirical and require mixing SI and English units.

k'_{20} = corrected treatability constant, $(gpm)^n/ft^2$
k_{20} = treatability constant = 0.075 $(gpm)^n/ft^2$
D = filter depth = 3.7 m (12 ft)
x = media coefficient = 0.3
θ = temperature correction coefficient = 1.06
T = temperature = $16°C$

$$k'_{20} = k_{20} \left(\frac{20 \text{ ft}}{D}\right)^x (\theta^{(T-20)})$$

$$= \left(0.075 \frac{(gpm)^n}{ft^2}\right) \left(\frac{20 \text{ ft}}{12 \text{ ft}}\right)^{0.3} (1.06^{(16-20)})$$

$$= 0.069 \text{ } (gpm)^n/ft^2$$

S_i = influent BOD with recirculated flow, mg/L
S = effluent BOD = 50 mg/L
S_0 = influent BOD = 434 mg/L
Q = flow rate = 1600 m^3/d
Q_R = recirculated flow rate

$$Q_R = (2)\left(1600 \frac{m^3}{d}\right) = 3200 \text{ m}^3/\text{d}$$

$$S_i = \frac{QS_0 + Q_R S}{Q + Q_R}$$

$$= \frac{\left(1600 \frac{m^3}{d}\right)\left(434 \frac{mg}{L}\right) + \left(3200 \frac{m^3}{d}\right)\left(50 \frac{mg}{L}\right)}{1600 \frac{m^3}{d} + 3200 \frac{m^3}{d}}$$

$$= 178 \text{ mg/L}$$

A_s = filter cross section surface area, ft^2
n = treatability constant coefficient = 0.5

$$A_s = Q \left(\frac{\ln\left(\frac{S}{S_i}\right)}{-k'_{20}D}\right)^{1/n}$$

$$= \frac{\left(1600 \frac{m^3}{d}\right)\left(264 \frac{gal}{m^3}\right)}{1440 \frac{min}{d}}$$

$$\times \left(\frac{\ln\left(\frac{50 \frac{mg}{L}}{178 \frac{mg}{L}}\right)}{\left(-0.069 \frac{(gpm)^n}{ft^2}\right)(12 \text{ ft})}\right)^{1/0.5}$$

$$= 690 \text{ ft}^2$$

$$A_s = (690 \text{ ft}^2)\left(\frac{1 \text{ m}}{3.28 \text{ ft}}\right)^2$$

$$= \boxed{64 \text{ m}^2}$$

The answer is A.

7.2. HLR = hydraulic loading rate

$$HLR = \frac{Q + Q_R}{A_s}$$

$$= \frac{1600 \frac{m^3}{d} + 3200 \frac{m^3}{d}}{\left(1440 \frac{min}{d}\right)(64 \text{ m}^2)}$$

$$= \boxed{0.052 \text{ m}^3/\text{m}^2\cdot\text{min}}$$

The answer is B.

7.3. OLR = organic loading rate

$$OLR = \frac{QS_0}{A_s D}$$

$$= \frac{\left(1600 \frac{m^3}{d}\right)\left(434 \frac{mg}{L}\right)\left(1000 \frac{L}{m^3}\right)}{(64 \text{ m}^2)(3.7 \text{ m})\left(10^6 \frac{mg}{kg}\right)}$$

$$= \boxed{2.9 \text{ kg/m}^3\cdot\text{d}}$$

The answer is C.

7.4. ω = distribution arm rotation speed, rpm
HLR = hydraulic loading rate
$\quad$ = 0.052 m^3/m^2·min
N = number of distribution arms = 2
DR = dosing rate = 20 cm/pass

$$\omega = \frac{HLR}{N(DR)}$$

$$= \frac{\left(0.052 \frac{m^3}{m^2\cdot min}\right)\left(100 \frac{cm}{m}\right)}{(2)\left(20 \frac{cm}{pass}\right)}$$

$$= \boxed{0.13 \text{ rpm}}$$

The answer is B.

SOLUTION 8

8.1. The solids loading rate is

$$\left(1600 \ \frac{mg}{L}\right) \left(8300 \ \frac{m^3}{d}\right) \left(10^{-6} \ \frac{kg}{mg}\right) \left(1000 \ \frac{L}{m^3}\right)$$

$$= \boxed{13\,280 \ kg/d \quad (13\,000 \ kg/d)}$$

The answer is C.

8.2. A_s = surface area based on solids flux, m²
Q = flow rate = 8300 m³/d
X = influent TSS = 1600 mg/L
G = solids flux = 2.6 kg/m²·h

$$A_s = \frac{QX}{G} = \frac{\left(8300 \ \frac{m^3}{d}\right)\left(1600 \ \frac{mg}{L}\right)}{\left(2.6 \ \frac{kg}{m^2 \cdot h}\right)\left(24 \ \frac{h}{d}\right)} \times \left(10^{-6} \ \frac{kg}{mg}\right)\left(1000 \ \frac{L}{m^3}\right)$$

$$= \boxed{213 \ m^2 \quad (210 \ m^2)}$$

The answer is B.

8.3. A_s = surface area based on settling velocity, m²
v_s = settling velocity = 1.29 m/h

$$A_s = \frac{Q}{v_s} = \frac{\left(8300 \ \frac{m^3}{d}\right)}{\left(1.29 \ \frac{m}{h}\right)\left(24 \ \frac{h}{d}\right)}$$

$$= \boxed{268 \ m^2 \quad (270 \ m^2)}$$

The answer is C.

8.4. The design surface area is the greater of the areas based on solids flux and settling velocity. The settling velocity area of 270 m² is greater than the solids flux area of 210 m².

The design area is $\boxed{270 \ m^2}$.

The answer is C.

8.5. q_o = overflow rate, m³/m²·h

$$q_o = \frac{Q}{A_s} = \frac{8300 \ \frac{m^3}{d}}{(270 \ m^2)\left(24 \ \frac{h}{d}\right)}$$

$$= \boxed{1.28 \ m^3/m^2 \cdot h \quad (1.3 \ m^3/m^2 \cdot h)}$$

The answer is B.

SOLUTION 9

The figure was prepared from the given settling data. Lines AB and BD are drawn from the linear sections of the settling curve.

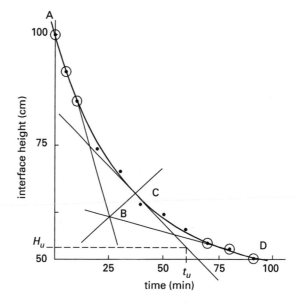

9.1. H_o = initial height of mixed suspension in settling column
  = 100 cm
C_o = concentration of TSS in mixed suspension
  = 5860 mg/L
C_u = concentration of TSS in thickened solids
  = 11 000 mg/L

$$H_u = \frac{C_o H_o}{C_u}$$

$$= \frac{\left(5860 \ \frac{mg}{L}\right)(100 \ cm)}{11\,000 \ \frac{mg}{L}}$$

$$= 53.3 \ cm \quad (53 \ cm)$$

A_t = surface area based on thickening, m²
Q = flow rate
  = 500 m³/d
t_u = settling time corresponding to point H_u on the figure
  = 63 min

$$A_t = \frac{Q t_u}{H_o}$$

$$= \frac{\left(500 \ \frac{m^3}{d}\right)(63 \ min)\left(100 \ \frac{cm}{m}\right)}{(100 \ cm)\left(1440 \ \frac{min}{d}\right)}$$

$$= \boxed{21.9 \ m^2 \quad (22 \ m^2)}$$

The answer is C.

9.2. From the figure, the negative slope of line AB gives the settling velocity, v_s.

$$v_s = \frac{-(100 \text{ cm} - 55 \text{ cm})\left(\frac{1 \text{ m}}{100 \text{ cm}}\right)}{(0 \text{ min} - 27 \text{ min})\left(\frac{1 \text{ h}}{60 \text{ min}}\right)}$$

$$= \boxed{1.0 \text{ m/h}}$$

The answer is C.

9.3. A_c = surface area based on settling velocity, m^2
Q_c = flow rate fractional liquid volume above the interface

$$Q_c = \frac{Q(H_o - H_u)}{H_o}$$

$$A_c = \frac{Q_c}{v_s} = \frac{\left(500 \frac{m^3}{d}\right)(100 \text{ cm} - 53 \text{ cm})}{(100 \text{ cm})\left(1.0 \frac{m}{h}\right)\left(24 \frac{h}{d}\right)}$$

$$= \boxed{9.8 \text{ m}^2}$$

The answer is B.

9.4. The design surface area, A_d, is the greater of A_t and A_c.

Since A_t is greater than A_c, $A_d = \boxed{22 \text{ m}^2}$.

The answer is C.

9.5. q_o = overflow rate

$$q_o = \frac{Q}{A_d} = \frac{500 \frac{m^3}{d}}{22 \text{ m}^2}$$

$$= \boxed{22.7 \text{ m}^3/\text{m}^2\text{·d} \quad (23 \text{ m}^3/\text{m}^2\text{·d})}$$

The answer is C.

9.6. V = solids volume, m^3/d
ρ_s = solids density, assume equal to water density
$= 1000 \text{ kg/m}^3$
f_s = solids fraction $= 18\,000 \text{ mg}/10^6 \text{ mg}$
$= 0.018$
m = daily mass flow of mixed suspension TSS, kg/d

$$V = \frac{m}{\rho_s f_s}$$

$$m = Q(\text{TSS})$$

$$= \left(500 \frac{m^3}{d}\right)\left(5860 \frac{mg}{L}\right)\left(10^{-6} \frac{kg}{mg}\right)\left(\frac{1000 \text{ L}}{m^3}\right)$$

$$= 2930 \text{ kg/d} \quad [\text{dry}]$$

$$V = \frac{2930 \frac{kg}{d}}{\left(1000 \frac{kg}{m^3}\right)(0.011)}$$

$$= \boxed{266 \text{ m}^3/\text{d} \quad (270 \text{ m}^3/\text{d})}$$

The answer is B.

SOLUTION 10

10.1. $CN^- + Cl_2 + H_2O \longrightarrow CNO^- + 2H^+ + 2Cl^-$

$$\frac{1400 \frac{mg}{L}}{\left(12 \frac{mg}{mmol} + 14 \frac{mg}{mmol}\right)} = 54 \text{ mmol/L CN}^-$$

1 mol CN^- reacts with 1 mol Cl_2. Therefore, 54 mmol/L CN^- will react with 54 mmol/L Cl_2.

$$\frac{\left(54 \frac{mmol}{L}\right)(2)\left(35.5 \frac{mg}{mmol}\right)}{\left(20\,000 \frac{m^3}{d}\right)\left(10^{-6} \frac{kg}{mg}\right)\left(1000 \frac{L}{m^3}\right)}$$

$$= \boxed{76\,680 \text{ kg/d} \quad (77\,000 \text{ kg/d})}$$

The answer is C.

10.2. $CN^- + O_3 \longrightarrow CNO^- + O_2$

1 mol CN^- reacts with 1 mol O_3. Therefore, 54 mmol/L CN^- will react with 54 mmol/L O_3.

$$\frac{\left(54 \frac{mmol}{L}\right)(3)\left(16 \frac{mg}{mmol}\right)}{\left(20\,000 \frac{m^3}{d}\right)\left(10^{-6} \frac{kg}{mg}\right)\left(1000 \frac{L}{m^3}\right)}$$

$$= \boxed{51\,840 \text{ kg/d} \quad (52\,000 \text{ kg/d})}$$

The answer is C.

10.3. The annual cost of chlorine is

$$\left(77\,000\ \frac{\text{kg}}{\text{d}}\right)\left(\frac{\$584}{1000\ \text{kg}}\right)\left(365\ \frac{\text{d}}{\text{yr}}\right)\left(\frac{100\%}{99.8\%}\right)$$

$$= \boxed{\$16{,}446{,}212/\text{yr} \quad (\$16{,}000{,}000/\text{yr})}$$

The answer is D.

10.4. The annual cost of ozone is

$$\left(52\,000\ \frac{\text{kg}}{\text{d}}\right)\left(14\ \frac{\text{kW·h}}{\text{kg}}\right)\left(\frac{\$0.042}{\text{kW·h}}\right)\left(365\ \frac{\text{d}}{\text{yr}}\right)$$

$$= \boxed{\$11{,}160{,}240/\text{yr} \quad (\$11{,}000{,}000/\text{yr})}$$

The answer is C.

SOLUTION 11

11.1. K_H = Henry's constant, unitless
ν_p = vapor pressure = 0.11 atm at 20°C
MW = molecular weight = 87 g/mol
S_w = solubility in water
   = 1250 mg/L at 20°C
R = universal gas constant
   = 8.2×10^{-5} atm·m^3/mol·K
T = temperature = 293K

$$K_H = \frac{\nu_p(\text{MW})}{S_w R T}$$

$$= \frac{(0.11\ \text{atm})\left(87\ \frac{\text{g}}{\text{mol}}\right)\left(1000\ \frac{\text{mg}}{\text{g}}\right)}{\left(1250\ \frac{\text{mg}}{\text{L}}\right)\left(8.2 \times 10^{-5}\ \frac{\text{atm·m}^3}{\text{mol·K}}\right)}$$
$$\times (293\text{K})\left(1000\ \frac{\text{L}}{\text{m}^3}\right)$$

$$= \boxed{0.32}$$

The answer is C.

11.2. V_a/V_w = air:water ratio, unitless
S = stripping factor = 4

$$\frac{V_a}{V_w} = \frac{S}{K_H} = \frac{4}{0.32} = \boxed{12.5 \quad (13)}$$

The answer is B.

11.3. HTU = transfer unit height, m
HLR$_m$ = mass hydraulic loading rate, kg/d
K_{La} = mass transfer coefficient = 0.019 s^{-1}
ρ_w = water density = 1000 kg/m^3

$$\text{HTU} = \frac{\text{HLR}_m}{K_{La}\rho_w}$$

$$= \frac{\left(3000\ \frac{\text{m}^3}{\text{d}}\right)\left(1000\ \frac{\text{kg}}{\text{m}^3}\right)}{(1\ \text{m}^2)\left(\frac{\pi}{4}\right)\left(86\,400\ \frac{\text{s}}{\text{d}}\right)}$$
$$\times (0.019\ \text{s}^{-1})\left(1000\ \frac{\text{kg}}{\text{m}^3}\right)$$

$$= \boxed{2.33\ \text{m} \quad (2.5\ \text{m})}$$

The answer is B.

11.4. NTU = number of transfer units
C_o = influent concentration, 1.57 mg/L
C = effluent concentration, 0.05 mg/L

$$\frac{C_o}{C} = \frac{1.57\ \frac{\text{mg}}{\text{L}}}{0.05\ \frac{\text{mg}}{\text{L}}} = 31.4$$

$$\text{NTU} = \left(\frac{S}{S-1}\right)\ln\left(\frac{\left(\frac{C_o}{C}\right)(S-1)+1}{S}\right)$$

$$= \left(\frac{4}{4-1}\right)\ln\left(\frac{(31.4)(4-1)+1}{4}\right)$$

$$= \boxed{4.23 \quad (4.5)}$$

The answer is D.

11.5. The packing height is

$$(\text{NTU})(\text{HTU}) = (4.5)(2.5\ \text{m}) = \boxed{11.25\ \text{m}}$$

The answer is C.

SOLUTION 12

period	average flow (m^3/period)	total flow per period (m^3)	cumulative flow (m^3)
0000–0200	275	550	550
0200–0400	389	778	1328
0400–0600	621	1242	2570
0600–0800	1340	2680	5250
0800–1000	1383	2766	8016
1000–1200	1312	2624	10 640
1200–1400	1098	2196	12 836
1400–1600	1027	2054	14 890
1600–1800	1084	2168	17 058
1800–2000	886	1772	18 830
2000–2200	259	518	19 348
2200–2400	326	652	20 000
		20 000	

$$\text{average flow} = \left(20\,000\ \frac{m^3}{d}\right)\left(\frac{1\ d}{24\ h}\right)$$
$$= 833\ m^3/h$$

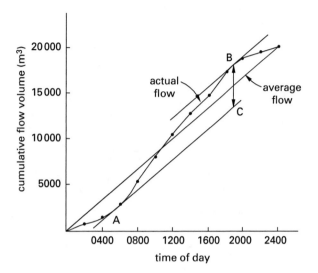

12.1. From the figure, the storage volume equals cumulative flow volume at point B minus cumulative flow volume at point C.

$$\text{storage volume} = 18\,000\ m^3 - 12\,500\ m^3$$
$$= \boxed{5500\ m^3}$$

The answer is B.

12.2. From the figure, the tank is filling whenever the slope of the actual flow line is greater than the slope of the average flow line.

The tank is full at point B on the figure corresponding to $\boxed{1900\ h.}$

The answer is C.

12.3. From the figure, the tank is emptying whenever the slope of the actual flow line is less than the slope of the average flow line.

The tank is empty at point A on the figure corresponding to $\boxed{0500\ h.}$

The answer is A.

SOLUTION 13

13.1.

land use	area (ha)	typical runoff coefficient
pasture	18	0.13
developed	1	0.75
woodland	$\underline{11}$	0.20
	30	

The weighted average runoff coefficient for the watershed is

$$\frac{(0.13)(18\ ha) + (0.75)(1\ ha) + (0.20)(11\ ha)}{30\ ha} = \boxed{0.18}$$

The answer is B.

13.2. t_c = time of concentration, min
C = runoff coefficient = 0.18
L = overland flow distance = 212 m = 696 ft
S = slope = 2.1%

$$t_c = \frac{(1.8)(1.1 - C)L^{1/2}}{S^{1/3}}$$
$$= \frac{(1.8)(1.1 - 0.18)(696\ ft)^{1/2}}{(2.1)^{1/3}}$$
$$= \boxed{34\ min}$$

The answer is B.

13.3. Q = flow, m^3/s
A = watershed area = 30 ha
i = rainfall intensity, cm/h

From the figure at $t_c = 34$ min, $i = 7.5$ cm/h.

$$Q = CiA$$
$$= (0.18)\left(7.5\ \frac{cm}{h}\right)(30\ ha)\left(\frac{1\ m}{100\ cm}\right)$$
$$\times \left(10\,000\ \frac{m^2}{ha}\right)\left(\frac{1\ h}{3600\ s}\right)$$
$$= \boxed{1.125\ m^3/s \quad (1.1\ m^3/s)}$$

The answer is B.

SOLUTION 14

14.1. The following figure shows a cross section of the ditch.

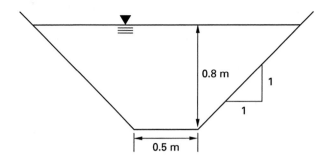

From the figure, the hydraulic radius of the ditch is

$$R = \frac{(0.8 \text{ m})^2 + (0.5 \text{ m})(0.8 \text{ m})}{0.5 \text{ m} + (2)(0.8^2 \text{ m}^2 + 0.8^2 \text{ m}^2)^{1/2}} = \frac{1.04 \text{ m}^2}{2.76 \text{ m}}$$

$$= 0.38 \text{ m}$$

v = flow velocity, m/s
n = Manning roughness coefficient
  = 0.025 (typical for earthen ditch)
S = slope, fraction = 1.8%/100% = 0.018

$$v = \frac{R^{2/3}S^{1/2}}{n}$$

$$= \frac{(0.38 \text{ m})^{2/3}(0.018)^{1/2}}{0.025}$$

$$= \boxed{2.8 \text{ m/s} \quad (3.0 \text{ m/s})}$$

The answer is B.

14.2. From the figure in Solution 1, the channel cross-sectional area is

$$A = (0.8 \text{ m})^2 + (0.5 \text{ m})(0.8 \text{ m}) = 1.04 \text{ m}^2$$

$$\text{flow rate} = Q = Av = (1.04 \text{ m}^2)\left(3.0 \ \frac{\text{m}}{\text{s}}\right)$$

$$= \boxed{3.12 \text{ m}^3/\text{s} \quad (3.0 \text{ m}^3/\text{s})}$$

The answer is B.

14.3.

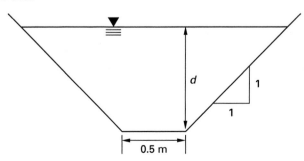

$$A = d(0.5 \text{ m} + d)$$

$$P = 0.5 \text{ m} + (2)(d^2 + d^2)^{1/2} = 0.5 \text{ m} + 2.83d$$

$$R = A/P$$

Q remains unchanged before and after lining with asphalt and the typical n for asphalt lining is 0.015.

$$Q = \frac{R^{2/3}S^{1/2}A}{n} = \frac{A^{5/3}S^{1/2}}{P^{2/3}n}$$

$$3.0 \ \frac{\text{m}^3}{\text{s}} = \frac{\left(d(0.5 \text{ m} + d)\right)^{5/3}(0.018)^{1/2}}{(0.5 \text{ m} + 2.83d)^{2/3}(0.015)}$$

Solve for d by trial and error.

$$d = \boxed{0.64 \text{ m} \quad (0.6 \text{ m})}$$

The answer is B.

SOLUTION 15

The figure is the synthesized hydrograph from the unit hydrographs given in the problem statement. It was constructed by summing the unit hydrograph runoff at 1 h increments.

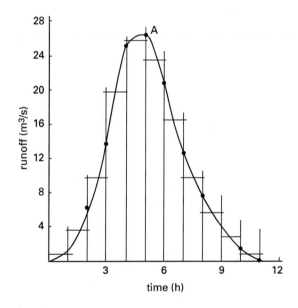

15.1. From the figure, the peak discharge occurs at point A and is $\boxed{26 \text{ m}^3/\text{s}.}$

The answer is D.

15.2. From the figure, the peak discharge occurs 5 h after the beginning of the storm.

The answer is B.

15.3. From the figure, the total runoff is equal to the area under the hydrograph curve. To find this area, integrate the hydrograph curve using 1 h time intervals.

runoff volume

$$= (1 \text{ h}) \left(3600 \, \frac{\text{s}}{\text{h}} \right)$$

$$\times \left(\begin{array}{c} 0.5 \, \dfrac{\text{m}^3}{\text{s}} + 3.5 \, \dfrac{\text{m}^3}{\text{s}} + 9.5 \, \dfrac{\text{m}^3}{\text{s}} \\[2ex] + 19.5 \, \dfrac{\text{m}^3}{\text{s}} + 25.5 \, \dfrac{\text{m}^3}{\text{s}} + 23.5 \, \dfrac{\text{m}^3}{\text{s}} \\[2ex] + 16.0 \, \dfrac{\text{m}^3}{\text{s}} + 9.5 \, \dfrac{\text{m}^3}{\text{s}} + 5.5 \, \dfrac{\text{m}^3}{\text{s}} \\[2ex] + 3.0 \, \dfrac{\text{m}^3}{\text{s}} + 0.5 \, \dfrac{\text{m}^3}{\text{s}} \end{array} \right)$$

$$= \boxed{419\,400 \text{ m}^3 \quad (420\,000 \text{ m}^3)}$$

The answer is D.

SOLUTION 16

The 20 yr of 24 h peak discharge values are ranked with the cumulative frequency in the table.

rank	annual peak discharge (m^3/s)	cumulative frequency
1	42	0.025
2	49	0.075
3	51	0.125
4	53	0.175
5	54	0.225
6	57	0.275
7	62	0.325
8	64	0.375
9	65	0.425
10	71	0.475
11	79	0.525
12	83	0.575
13	88	0.625
14	89	0.675
15	92	0.725
16	94	0.775
17	96	0.825
18	103	0.875
19	112	0.925
20	128	0.975

$$\text{cumulative frequency} = \frac{\text{rank} - 0.5}{20}$$

Solution 16

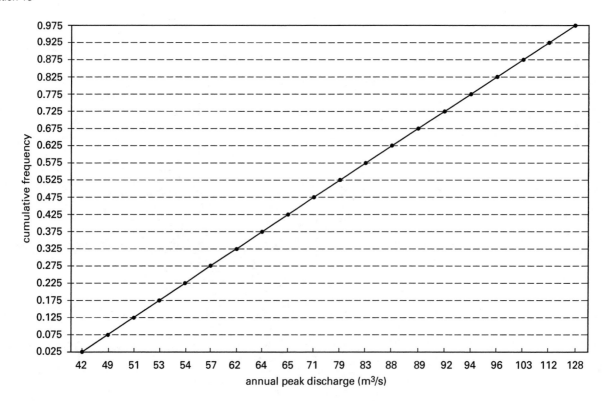

16.1. The 2 yr storm recurs at 2/20 or 0.1 frequency.

From the table or figure, a 0.1 frequency corresponds approximately to a peak discharge of 50 m³/s.

2 yr, 24 h storm peak discharge = $\boxed{50 \text{ m}^3/\text{s}}$

The answer is C.

16.2. The 10 yr storm recurs at 10/20 or 0.5 frequency.

From the table or figure, a 0.5 frequency corresponds approximately to a peak discharge of 75 m³/s.

10 yr, 24 h storm peak discharge = $\boxed{75 \text{ m}^3/\text{s}}$

The answer is B.

SOLUTION 17

17.1. k = intrinsic permeability, m²
K = hydraulic conductivity with water as
the fluid
= 0.42 m/d
μ = dynamic viscosity
= 1.39×10^{-3} kg/m s for water at 8°C
ρ = fluid density = 1000 kg/m³ for water
g = gravitational constant = 9.81 m/s²

$$
k = \frac{K\mu}{\rho g}
$$

$$
= \frac{\left(0.42 \, \frac{\text{m}}{\text{d}}\right)\left(1.39 \times 10^{-3} \, \frac{\text{kg}}{\text{m·s}}\right)\left(\frac{1 \text{ d}}{86\,400 \text{ s}}\right)}{\left(1000 \, \frac{\text{kg}}{\text{m}^3}\right)\left(9.81 \, \frac{\text{m}}{\text{s}^2}\right)}
$$

$$
= \boxed{6.9 \times 10^{-13} \text{ m}^2}
$$

The answer is D.

17.2. K = hydraulic conductivity for no. 2 fuel oil,
m/d
k = intrinsic permeability = 6.9×10^{-13} m²
ρ = fluid density
= 900 kg/m³ for no. 2 fuel oil
μ = dynamic viscosity
= 6.5×10^{-3} kg/m·s for no. 2 fuel oil at 8°C

$$
K = \frac{k\rho g}{\mu}
$$

$$
= \frac{(6.9 \times 10^{-13} \text{ m}^2)\left(900 \, \frac{\text{kg}}{\text{m}^3}\right)\left(9.81 \, \frac{\text{m}}{\text{s}^2}\right)}{\left(6.5 \times 10^{-3} \, \frac{\text{kg}}{\text{m·s}}\right)\left(\frac{1 \text{ d}}{86\,400 \text{ s}}\right)}
$$

$$
= \boxed{0.081 \text{ m/d}}
$$

The answer is C.

17.3. v_f = fuel velocity, m/d
K = hydraulic conductivity for no. 2 fuel oil
= 0.081 m/d
i = gradient = 0.022
n_e = effective porosity = 0.28

$$
v_f = \frac{Ki}{n_e}
$$

$$
= \frac{\left(0.081 \, \frac{\text{m}}{\text{d}}\right)(0.022)}{0.28}
$$

$$
= \boxed{6.4 \times 10^{-3} \text{ m/d}}
$$

The answer is B.

SOLUTION 18

18.1. Check each reaction order equation against the data in the table to see which equation plots a straight line.

elapsed time of aeration (s)	target VOC effluent concentration (μg/L)
0	990
30	497
60	251
90	124
120	63
180	14
240	3

Check zero-order.

$$
C = -kt
$$

$$
-\frac{C}{t} = k
$$

$$
\frac{497}{30} = 16.6
$$

$$
\frac{251}{60} = 4.2
$$

Because k is not constant, the reaction is not zero-order.

Check first-order.

$$\ln \frac{C}{C_o} = -kt$$

$$\frac{\ln \frac{C}{C_o}}{-t} = k$$

$$\frac{-\ln \left(\frac{497}{990} \right)}{30} = 0.0230$$

$$\frac{-\ln \left(\frac{251}{990} \right)}{60} = 0.0229$$

Because k is approximately constant, the reaction is probably first-order.

Check second-order.

$$\frac{1}{C} = kt + \frac{1}{C_o}$$

$$\frac{\frac{1}{C} - \frac{1}{C_o}}{t} = k$$

$$\frac{\frac{1}{497} - \frac{1}{990}}{30} = 3.34 \times 10^{-5}$$

$$\frac{\frac{1}{251} - \frac{1}{990}}{60} = 4.96 \times 10^{-5}$$

Because k is not constant, the reaction is not second-order.

The reaction is first-order.

The answer is B.

18.2. K_{La} = mass transfer coefficient, s^{-1}
C_o = initial concentration = 990 μg/L
C = concentration at each time increment, μg/L
t = time, s

$$K_{La} = \frac{-\ln \left(\frac{C}{C_o} \right)}{t}$$

This equation applies because the reaction is first-order.

t (s)	C (μg/L)	$\frac{-\ln(C/C_o)}{t}$
30	497	0.0230
60	251	0.0229
90	124	0.0231
120	63	0.0230
180	14	0.0237
240	3	0.0242
		average = 0.0233 s^{-1}

$$K_{La} = \boxed{0.023 \text{ s}^{-1}}$$

The answer is A.

18.3.

$$t = \frac{-\ln \left(\frac{1.0 \frac{\mu g}{L}}{990 \frac{\mu g}{L}} \right)}{0.023 \text{ s}^{-1}} = \boxed{300 \text{ s}}$$

The answer is B.

SOLUTION 19

19.1. The final dissolved oxygen in bottle 1 is greater than 7.0 mg/L, and in bottle 5, it is less than 2.0 mg/L. Therefore, both of these bottles are excluded from the calculations. Assume standard 300 mL BOD bottles are used.

$$\text{BOD bottle 2} = \frac{9.2 \frac{\text{mg}}{\text{L}} - 6.6 \frac{\text{mg}}{\text{L}}}{\frac{10 \text{ mL}}{300 \text{ mL}}} = 78 \text{ mg/L}$$

$$\text{BOD bottle 3} = \frac{9.1 \frac{\text{mg}}{\text{L}} - 5.2 \frac{\text{mg}}{\text{L}}}{\frac{15 \text{ mL}}{300 \text{ mL}}} = 78 \text{ mg/L}$$

$$\text{BOD bottle 4} = \frac{9.1 \frac{\text{mg}}{\text{L}} - 4.1 \frac{\text{mg}}{\text{L}}}{\frac{20 \text{ mL}}{300 \text{ mL}}} = 75 \text{ mg/L}$$

$$\text{BOD}_5 \text{ at } 20 \,°\text{C} = \frac{78 \frac{\text{mg}}{\text{L}} + 78 \frac{\text{mg}}{\text{L}} + 75 \frac{\text{mg}}{\text{L}}}{3}$$

$$= \boxed{77 \text{ mg/L}}$$

The answer is B.

19.2. BOD_u = ultimate BOD, mg/L
t = time, d
BOD_t = BOD at some time, t, mg/L
k = rate coefficient = 0.40 d^{-1} at 20°C

$$\text{BOD}_u = \frac{\text{BOD}_t}{1 - e^{-kt}}$$

$$= \frac{77 \frac{\text{mg}}{\text{L}}}{1 - e^{(-0.40 \text{ d}^{-1})(5 \text{ d})}}$$

$$= \boxed{89 \text{ mg/L}}$$

The answer is B.

19.3. k_{15} = rate coefficient at 15°C, d^{-1}
k_{20} = rate coefficient at 20°C
$= 0.40$ d^{-1}
θ = temperature correction coefficient, unitless
$= 1.047$

$$k_{15} = k_{20}\theta^{15-20}$$
$$= \left(0.40 \text{ d}^{-1}\right)\left(1.047^{(15-20)}\right)$$
$$= 0.32 \text{ d}^{-1}$$

$$\text{BOD}_7 \text{ at } 15°\text{C} = \left(89 \; \frac{\text{mg}}{\text{L}}\right)\left(1 - e^{(-0.32 \text{ d}^{-1})(7 \text{ d})}\right)$$

$$= \boxed{79.5 \text{ mg/L } (80 \text{ mg/L})}$$

The answer is B.

Potable Water

PROBLEM 1

Bench studies have determined that aluminum sulfate (alum) promotes acceptable floc formation at a dose of 23 mg/L when applied to a surface water source. Design standards for chemical flocculation require flash mixing at a minimum velocity gradient (G) of 900/s for 120 seconds. Alum is available at 17% purity for $234/1000 kg and electrical power costs $0.05/kW·h. The water demand is 19 000 m³/d. The water temperature is 20°C.

1.1. What is the total tank volume required for flash mixing?

(A) 2.7 m³
(B) 16 m³
(C) 27 m³
(D) 160 m³

1.2. If four tanks are desired, what are the dimensions of each tank?

(A) $l = 5.4$ m, $w = 1.9$ m, $d = 3.0$ m
(B) $l = 1.9$ m, $w = 1.9$ m, $d = 1.9$ m
(C) $l = 3.0$ m, $w = 3.0$ m, $d = 3.0$ m
(D) $l = 5.4$ m, $w = 5.4$ m, $d = 5.4$ m

1.3. What is the total power required for flash mixing if the motor efficiency is 85%?

(A) 2.2×10^3 N·m/s
(B) 1.5×10^4 N·m/s
(C) 2.6×10^4 N·m/s
(D) 1.3×10^5 N·m/s

1.4. What is the total monthly cost for alum?

(A) $600/month
(B) $3100/month
(C) $18,000/month
(D) $180,000/month

1.5. What is the total monthly cost for electricity if the motor efficiency is 85%?

(A) $80/month
(B) $470/month
(C) $790/month
(D) $940/month

PROBLEM 2

Flocculation tanks need to be sized to match three sedimentation basins 3.5 m deep and 8.0 m wide. The sedimentation basins each handle a flow of 4000 m³/d, as will the flocculation tanks. The flocculation tanks will have two mixing sections with an overall average velocity gradient (G) of 30/s. The average velocity gradient-time (Gt) value has been set at 89 700. The paddles will be the wooden paddle-wheel type and will turn on a horizontal axis perpendicular to flow, with one paddle wheel in each tank section. A common motor will turn the paddles in both sections of each tank at a constant rotational speed of 3 rpm. The water temperature is 20°C.

2.1. What are the flocculation tank dimensions?

(A) $l = 4.0$ m, $w = 4.0$ m, $d = 4.0$ m
(B) $l = 5.2$ m, $w = 5.2$ m, $d = 5.2$ m
(C) $l = 5.0$ m, $w = 8.0$ m, $d = 3.5$ m
(D) $l = 7.0$ m, $w = 6.0$ m, $d = 3.5$ m

2.2. What are the G values for the first (inlet end) and for the second (outlet end) sections of each flocculation tank?

(A) $G_1 = 15/s$, $G_2 = 45/s$
(B) $G_1 = 20/s$, $G_2 = 40/s$
(C) $G_1 = 30/s$, $G_2 = 30/s$
(D) $G_1 = 40/s$, $G_2 = 20/s$

2.3. What are the power requirements for the paddles in the first and second sections of each flocculation tank?

(A) $P_1 = 3.0$ N·m/s, $P_2 = 1.5$ N·m/s
(B) $P_1 = 66$ N·m/s, $P_2 = 66$ N·m/s
(C) $P_1 = 120$ N·m/s, $P_2 = 30$ N·m/s
(D) $P_1 = 240$ N·m/s, $P_2 = 60$ N·m/s

2.4. What are the required paddle areas for the first and second sections of each flocculation tank?

(A) $A_1 = 0.19$ m^2, $A_2 = 0.048$ m^2
(B) $A_1 = 0.39$ m^2, $A_2 = 0.10$ m^2
(C) $A_1 = 0.84$ m^2, $A_2 = 0.21$ m^2
(D) $A_1 = 3.2$ m^2, $A_2 = 0.81$ m^2

PROBLEM 3

A treatment plant upgrade calls for design of sedimentation basins to handle a flow rate of 30 000 m^3/d. The basins are for a Type I suspension with an overflow rate (q_o) of 1.1 m^3/m^2·h, a length to width ratio of 4:1, a weir overflow rate (q_w) of 14 m^3/m·h, and a maximum settling zone length of 40 m. The minimum settling zone depth (Z_o) is 2.5 m and 0.5 meters is allowed for freeboard.

3.1. What is the minimum number of tanks required to treat the flow?

(A) 1
(B) 2
(C) 3
(D) 4

3.2. What are the surface dimensions of the settling zone for each tank?

(A) $l = 68$ m, $w = 17$ m
(B) $l = 48$ m, $w = 12$ m
(C) $l = 40$ m, $w = 10$ m
(D) $l = 34$ m, $w = 8.5$ m

3.3. What are the total length, width, and depth dimensions for each tank?

(A) $l = 40$ m, $w = 10$ m, $d = 3.0$ m
(B) $l = 45$ m, $w = 10$ m, $d = 3.5$ m
(C) $l = 48$ m, $w = 12$ m, $d = 3.5$ m
(D) $l = 68$ m, $w = 17$ m, $d = 3.5$ m

3.4. What is the required weir length for each tank?

(A) 15 m
(B) 30 m
(C) 45 m
(D) 90 m

PROBLEM 4

The chlorination curve in the figure was prepared from representative samples collected following filtration at a municipal water treatment plant.

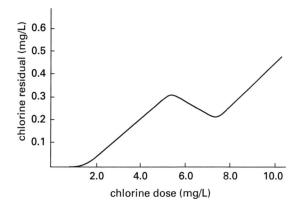

4.1. At what dose does breakpoint chlorination begin?

(A) 0.2 mg/L
(B) 1.4 mg/L
(C) 4.9 mg/L
(D) 7.5 mg/L

4.2. What dose is required to produce a free chlorine residual of 0.4 mg/L?

(A) 1.4 mg/L
(B) 4.9 mg/L
(C) 7.5 mg/L
(D) 9.6 mg/L

PROBLEM 5

Column tests have produced results presented in the following table that describe the settling characteristics of a Type II suspension. The initial TSS concentration of the completely mixed suspension was 208 mg/L. The suspension will be treated for TSS removal at a flow rate of 2000 m^3/d.

time (min)	percent TSS removed at indicated depth (m)					
	1.0	1.5	2.0	2.5	3.0	3.5
80	72	68	64	59	53	46
90	76	72	67	64	57	51
100	79	76	72	66	62	54
110	83	77	73	70	64	58
120	87	83	77	74	68	65
130	92	89	84	80	74	69
140	96	93	88	85	82	79

5.1. What is the residence time required to achieve 85% TSS removal if the settling zone depth is 3.0 m?

(A) 110 min
(B) 120 min
(C) 130 min
(D) 140 min

5.2. What is the volume of the settling zone?

(A) 160 m^3
(B) 170 m^3
(C) 180 m^3
(D) 200 m^3

5.3. What is the surface area of the settling zone?

(A) 52 m^2
(B) 57 m^2
(C) 60 m^2
(D) 65 m^2

PROBLEM 6

A small municipality wants to apply direct filtration to improve the quality of its drinking water. The water is supplied through three groundwater wells providing a total flow of 3000 m^3/d. A uniform media filter has been selected for the project. Filter media and operating characteristics are as follows.

media packed bed porosity = 0.43

media fluidized bed porosity = 0.65

media mean particle diameter = 0.68 mm

media shape factor = 0.82

media particle settling velocity = 0.10 m/s

packed bed depth = 0.75 m

clean bed filtering velocity = 4.6 m/h

water temperature = 8°C

6.1. What is the head loss at the beginning of the filter run?

(A) 0.042 m
(B) 0.27 m
(C) 3.1 m
(D) 24 m

6.2. What is the filter bed depth during backwash?

(A) 0.86 m
(B) 1.2 m
(C) 9.1 m
(D) 10 m

6.3. What is the backwash velocity?

(A) 0.0078 m^3/m^2·min
(B) 0.094 m^3/m^2·min
(C) 0.84 m^3/m^2·min
(D) 6.7 m^3/m^2·min

6.4. What is the filter bed total surface area?

(A) 2.5 m^2
(B) 14 m^2
(C) 27 m^2
(D) 650 m^2

PROBLEM 7

A municipal drinking water service authority is considering adding softening to its water treatment. The facility being evaluated processes a flow rate of 18 000 m^3/d with the following hardness ion distribution.

Ca^{+2} = 347 mg/L

Mg^{+2} = 129 mg/L

HCO_3^- = 1256 mg/L

Softening chemicals are available at the following purity and cost.

CaO 67% purity at $82/1000 kg
Na_2CO_3 98% purity at $105/1000 kg
NaOH 73% purity at $215/1000 kg

7.1. What is the annual cost for lime and soda ash to be used as reagents in softening of the water, assuming 100% removal of the hardness?

(A) $230,000/yr
(B) $450,000/yr
(C) $620,000/yr
(D) $990,000/yr

7.2. What is the annual dry sludge mass generated from lime-soda ash softening of the water, assuming 100% removal of the hardness?

(A) 1.6×10^5 kg/yr (dry)

(B) 2.3×10^6 kg/yr (dry)

(C) 1.8×10^7 kg/yr (dry)

(D) 2.1×10^8 kg/yr (dry)

PROBLEM 8

A commercial laundry is considering constructing an ion exchange process to remove hardness from the water used in laundering operations. The laundry uses 1000 m³/d of city water with Ca^{+2} at 4.5 meq/L and Mg^{+2} at 2.0 meq/L. Their goal is to reduce total hardness (TH) to 1.0 meq/L. The softening resin has a TH capacity of 60 kg TH/m³ of media with a recommended hydraulic loading rate of 0.4 m³/m²·min. Regeneration can be accomplished using 100 kg NaCl/m³ in a 15% solution at a hydraulic loading rate of 0.04 m³/m²·min. The exchanger tanks available to the laundry have a diameter of 1.0 m.

8.1. What is the total bypass-water flow rate?

(A) 0 m³/d

(B) 150 m³/d

(C) 850 m³/d

(D) 1000 m³/d

8.2. What is the total daily resin volume required to soften the water?

(A) 4.6 m³/d

(B) 5.4 m³/d

(C) 280 m³/d

(D) 330 m³/d

8.3. How many exchanger tanks are required?

(A) 1

(B) 2

(C) 3

(D) 4

8.4. What is the resin bed depth in each exchanger tank?

(A) 1.5 m

(B) 1.9 m

(C) 2.9 m

(D) 5.8 m

8.5. What is the daily volume of NaCl regeneration solution used per tank?

(A) 0.78 m³/d

(B) 1.5 m³/d

(C) 1.9 m³/d

(D) 3.8 m³/d

8.6. What is the time period required for resin regeneration in any single tank?

(A) 26 min

(B) 39 min

(C) 48 min

(D) 98 min

PROBLEM 9

A groundwater extraction system produces a flow of 2.0 m³/min containing a mixture of the following five organic chemicals.

chemical	concentration (mg/L)
1,1,1-trichloroethane	2.90
trichloroethylene	1.40
1,1-dichloroethane	0.11
1,1-dichloroethylene	0.32
methylene chloride	0.61

Bench scale adsorption isotherm tests were conducted using the chemical mixture. The resulting isotherm equation is

$$\frac{x}{M} = 5.37 C_r^{0.58}$$

x is the chemical mass removed and M is the GAC mass used.

The summed concentration of the five chemicals must be reduced to 0.005 mg/L in order to satisfy discharge requirements. To ease operational requirements, the granular activated carbon (GAC) change-out period should be should between 60 and 90 days.

9.1. What is the activated carbon use rate?

(A) 4.0 kg/d

(B) 7.0 kg/d

(C) 35 kg/d

(D) 62 kg/d

9.2. What is the adsorption vessel size, using standard vessel sizes?

(A) 900 kg GAC

(B) 1800 kg GAC

(C) 4500 kg GAC

(D) 9000 kg GAC

9.3. What is the GAC change-out period?

(A) 51 d

(B) 60 d

(C) 73 d

(D) 90 d

9.4. What is the empty bed contact time?

(A) 2 min

(B) 5 min

(C) 10 min

(D) 12 min

PROBLEM 10

A chemical analysis of a water sample has produced the results presented in the following table.

cation	concentration (mg/L)	anion	concentration (mg/L)
Ca^{+2}	158	SO_4^{-2}	64
Mg^{+2}	47	Cl^-	43
Na^+	26	HCO_3^-	381
K^+	19		

10.1. Is the analysis complete?

(A) No, the cation concentration in meq/L exceeds the anion concentration in meq/L by more than 10%.

(B) Yes, the cation concentration in meq/L exceeds the anion concentration in meq/L by more than 10%.

(C) No, the cation concentration in mg/L exceeds the anion concentration in mg/L by more than 10%.

(D) Yes, the cation concentration in mg/L exceeds the anion concentration in mg/L by more than 10%.

10.2. What is the total hardness?

(A) 210 mg/L

(B) 390 mg/L as $CaCO_3$

(C) 590 mg/L as $CaCO_3$

(D) 740 mg/L

10.3. What is the carbonate hardness?

(A) 210 mg/L

(B) 310 mg/L as $CaCO_3$

(C) 380 mg/L

(D) 590 mg/L as $CaCO_3$

10.4. What is the non-carbonate hardness?

(A) 210 mg/L

(B) 280 mg/L as $CaCO_3$

(C) 310 mg/L as $CaCO_3$

(D) 590 mg/L as $CaCO_3$

PROBLEM 11

Chlorine gas is used for disinfection of a municipal drinking water supply. The chlorine gas reacts with the water to form hypochlorous acid at a concentration of 4 mg/L. The water temperature is 15°C. The pK_a for hypochlorous acid at 15°C is 7.63.

11.1. What is the hydrogen ion concentration in the water from the dissociation of the hypochlorous acid?

(A) 2.3×10^{-8} M

(B) 1.3×10^{-6} M

(C) 7.5×10^{-5} M

(D) 7.6×10^{-5} M

11.2. What is the hypochlorite concentration in the solution from the dissociation of the hypochlorous acid?

(A) 2.3×10^{-8} M

(B) 1.3×10^{-6} M

(C) 7.5×10^{-5} M

(D) 7.6×10^{-5} M

11.3. What is the percent ionization of the hypochlorous acid in the water?

(A) 1.3%

(B) 13%

(C) 74%

(D) 100%

PROBLEM 12

A water treated for manufacturing uses will be distributed through a network of steel pipes. The water, with a temperature of 25°C and a pH of 7.4, is characterized by the following ionic constituents.

ion	concentration (mg/L)
Ca^{+2}	46
Na^+	128
Cl^-	197
HCO_3^-	133

Assume the ionic strength and temperature constant is 2.28.

12.1. What is the ionic strength of the water?

(A)   0.0078

(B)   0.0090

(C)   0.016

(D)   0.018

12.2. What is the Langelier saturation index for the water?

(A)   −0.60

(B)   −0.48

(C)   0.15

(D)   0.43

12.3. Is the water corrosive to iron pipes?

(A)   Yes, the water is corrosive because SI is negative.

(B)   Yes, the water is corrosive because SI is positive.

(C)   No, the water is noncorrosive because SI is negative.

(D)   No, the water is noncorrosive because SI is positive.

PROBLEM 13

A city's census records show the population growth presented in the following table.

year	population
1950	12 200
1960	18 000
1970	23 500
1980	30 000
1990	38 600

13.1. What is the projected population for the city in 2030?

(A)   48 000 people

(B)   63 000 people

(C)   100 000 people

(D)   122 000 people

13.2. If the city's water source is groundwater, what is an appropriate design life for groundwater wells?

(A)   5 to 10 yr

(B)   10 to 20 yr

(C)   20 to 40 yr

(D)   over 40 yr

13.3. If the city's water source is surface water, what is an appropriate design life for a water treatment plant?

(A)   5 to 10 yr

(B)   10 to 20 yr

(C)   20 to 40 yr

(D)   over 40 yr

13.4. What is an appropriate design life for pumping facilities to service the city's water distribution system reservoirs?

(A)   5 to 10 yr

(B)   10 to 20 yr

(C)   20 to 40 yr

(D)   over 40 yr

PROBLEM 14

The figure presents the projected 24 h flow distribution for a small municipality. The municipality currently has the capacity to pump at 3.6 m^3/min on a continuous basis and is considering adding a storage tank instead of increasing pumping capacity to meet the projected demand.

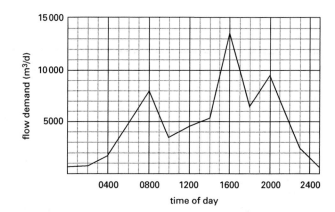

14.1. What is the maximum pumping rate required to meet peak demand if no storage is provided?

(A)   0.040 m^3/s

(B)   0.16 m^3/s

(C)   0.40 m^3/s

(D)   1.6 m^3/s

14.2. What is the required storage volume to meet a uniform 24 h pumping rate of 3.6 m^3/min?

(A)   1250 m^3

(B)   5200 m^3

(C)   14 000 m^3

(D)   34 000 m^3

PROBLEM 15

A pipe network is presented in the figure. The values of C and n for the Hazen-Williams equation are 100 and 1.85, respectively.

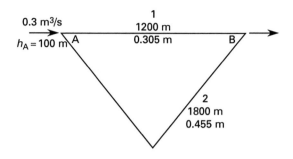

15.1. What is the head at node B?

(A) 76 m

(B) 85 m

(C) 90 m

(D) 100 m

15.2. What are the flows in pipes 1 and 2?

(A) $Q_1 = 0.10$ m^3/s, $Q_2 = 0.20$ m^3/s

(B) $Q_1 = 0.15$ m^3/s, $Q_2 = 0.15$ m^3/s

(C) $Q_1 = 0.20$ m^3/s, $Q_2 = 0.10$ m^3/s

(D) $Q_1 = 0.28$ m^3/s, $Q_2 = 0.02$ m^3/s

PROBLEM 16

Benzene, toluene, and xylene have been discovered in the groundwater of an unconfined aquifer. The effective porosity and bulk density of the aquifer soil are 0.34 and 1.83 g/cm^3, respectively. The soil TOC concentration is 148 mg/kg.

16.1. What are the maximum possible benzene, toluene, and xylene solute concentrations at or near the release point?

(A) 5 μg/L, 1000 μg/L, 10 000 μg/L

(B) 1800 mg/L, 500 mg/L, 160 mg/L

(C) no maximum concentration limits exist

(D) unable to determine from the information provided

16.2. Which of the three chemicals will move away from the source at the highest velocity?

(A) benzene

(B) toluene

(C) xylene

(D) all will move at the same velocity

16.3. What is the relative velocity of benzene to groundwater?

(A) 1.1

(B) 1.0

(C) 0.93

(D) 0.077

16.4. If the groundwater velocity is 0.23 m/d, how long will it take for the center of mass of the benzene plume to move 100 m from the source?

(A) 400 d

(B) 430 d

(C) 480 d

(D) 5600 d

PROBLEM 17

Leaks have been discovered in piping from an underground tank used to store tetrachloroethylene (PCE) at a commercial dry cleaning business. It is believed that the release has been essentially continuous for the past 2 yr and that approximately 30 m^3 of the PCE has been released during that period. The site's hydrogeologic setting is characterized by an unconfined aquifer with an average groundwater velocity of 0.37 m/d and a retardation factor for PCE of 1.08. The PCE concentration at the source is 1212 μg/L.

17.1. What is the average velocity of the PCE plume where $C/C_o = 0.5$?

(A) 0.30 m/d

(B) 0.34 m/d

(C) 0.37 m/d

(D) 0.40 m/d

17.2. How long will or did it take for the PCE to reach the property boundary 50 m downgradient from the release point at a concentration equal to its maximum contaminant level (MCL)?

(A) 47 d

(B) 80 d

(C) 270 d

(D) 600 d

Potable Water Solutions

SOLUTION 1

1.1. V = mixing volume, m³
Q = flow rate = 19 000 m³/d
t = detention time = 120 s

$$V = Qt$$

$$= \frac{\left(19\,000\,\frac{\text{m}^3}{\text{d}}\right) 120\text{ s}}{86\,400\,\frac{\text{s}}{\text{d}}}$$

$$= \boxed{26.4\text{ m}^3 \quad (27\text{ m}^3)}$$

The answer is C.

1.2. $$\frac{27\text{ m}^3}{4\text{ tanks}} = 6.75\text{ m}^3/\text{tank}$$

For mixing, cubic dimensions are desired. Therefore, $l = w = d$ and $V = w^3$.

$$w = (6.75\text{m}^3)^{1/3}$$

$$= \boxed{1.9\text{ m}}$$

$$l = \boxed{1.9\text{ m}}$$

$$d = \boxed{1.9\text{ m}}$$

The answer is B.

1.3. P = power, N·m/s
G = velocity gradient = 900/s
μ = dynamic viscosity = 1.002×10^{-3} N·s/m²

$$P = G^2 V \mu$$

$$= \left(\frac{900}{\text{s}}\right)^2 (27\text{ m}^3)\left(1.002 \times 10^{-3}\,\frac{\text{N·s}}{\text{m}^2}\right)$$

$$= 21\,914\text{ N·m/s}$$

P at 85% efficiency

$$= \frac{21\,914\,\frac{\text{N·m}}{\text{s}}}{0.85}$$

$$= \boxed{25\,780\text{ N·m/s} \quad (2.6 \times 10^4\text{ N·m/s})}$$

The answer is C.

1.4. $\left(19\,000\,\frac{\text{m}^3}{\text{d}}\right)\left(23\,\frac{\text{mg}}{\text{L}}\right)\left(1000\,\frac{\text{L}}{\text{m}^3}\right)$

$$\times \left(30\,\frac{\text{d}}{\text{mo}}\right)\left(10^{-6}\,\frac{\text{kg}}{\text{mg}}\right)$$

$$= 13\,110\text{ kg/mo}$$

$$\left(13\,110\,\frac{\text{kg}}{\text{mo}}\right)\left(\frac{\$234}{1000\text{ kg}}\right)\left(\frac{100\%}{17\%}\right)$$

$$= \boxed{\$18,046/\text{mo} \quad (\$18,000/\text{mo})}$$

The answer is C.

1.5.

$$\frac{\left(2.6 \times 10^4\,\frac{\text{N·m}}{\text{s}}\right)\left(\frac{\$0.05}{\text{kW·h}}\right)\left(30\,\frac{\text{d}}{\text{mo}}\right)\left(24\,\frac{\text{h}}{\text{d}}\right)\text{kW}}{1000\,\frac{\text{N·m}}{\text{s}}}$$

$$= \boxed{\$936/\text{mo} \quad (\$940/\text{mo})}$$

The answer is D.

SOLUTION 2

2.1. t = detention time, s
Gt = time-velocity gradient, unitless = 89 700
G = velocity gradient, 30 s⁻¹

$$t = \frac{Gt}{G}$$

$$= \frac{89\,700}{\left(\frac{30}{\text{s}}\right)\left(60\,\frac{\text{s}}{\text{min}}\right)}$$

$$= 50\text{ min}$$

V = tank volume, m^3
Q = flow rate = 4000 m^3/d

$$V = Qt$$

$$= \frac{\left(4000 \; \dfrac{m^3}{d}\right)(50 \text{ min})}{1440 \; \dfrac{\text{min}}{d}}$$

$$= 139 \text{ m}^3$$

To match the sedimentation basin, the flocculation tank depth should equal the sedimentation basin depth of 3.5 m and the maximum flocculation basin width should be equal to the sedimentation basin width of 8.0 m.

For the most efficient mixing with wooden paddle-wheel type paddles turning on a horizontal axis perpendicular to flow, each mixing section should have square dimensions along the cross section of the tank length. A minimum clearance of 0.3 m should be provided between the paddle tip and the tank walls and floors to avoid floc shear.

Applying these criteria, the flocculation tank dimensions are

$$d = \boxed{3.5 \text{ m}}$$

$$l = \boxed{7.0 \text{ m}}$$

$$w = \frac{139 \text{ m}^3}{(3.5 \text{ m})(7.0 \text{ m})}$$

$$= \boxed{5.7 \text{ m} \quad (6.0 \text{ m})} \quad [< 8.0 \text{ m, therefore OK}]$$

The following figure illustrates the tank cross section.

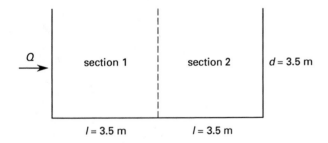

$l = 3.5$ m $l = 3.5$ m

The answer is D.

2.2. $$G_{\text{ave}} = 30/s$$

For good floc formation in a two-section tank, G_1 at the inlet should be greater than G_2 at the outlet. Let $G_1 = 2G_2$.

$$\frac{G_1 + G_2}{2} = \frac{2G_2 + G_2}{2} = 30/s$$

$$3G_2 = 60/s$$

$$G_2 = \boxed{20/s}$$

$$2G_2 = G_1 = \boxed{40/s}$$

The answer is D.

2.3. P = power, N·m/s
G = velocity gradient for each section, s^{-1}
V = volume, m^3
μ = dynamic viscosity = 1.002×10^{-3} N·s/m^2

$$V = \left(\frac{7.0 \text{ m}}{2 \text{ sections}}\right)(6.0 \text{ m})(3.5 \text{ m}) = 73.5 \text{ m}^3$$

$$P = G^2 V \mu$$

$$P_1 = \left(\frac{40}{s}\right)^2 (73.5 \text{ m}^3)\left(1.002 \times 10^{-3} \; \frac{\text{N·s}}{\text{m}^2}\right)$$

$$= \boxed{118 \text{ N·m/s} \quad (120 \text{ N·m/s})}$$

$$P_2 = \left(\frac{20}{s}\right)^2 (73.5 \text{ m}^3)\left(1.002 \times 10^{-3} \; \frac{\text{N·s}}{\text{m}^2}\right)$$

$$= \boxed{29.5 \text{ N·m/s} \quad (30 \text{ N·m/s})}$$

The answer is C.

2.4. v = paddle velocity, m/s
v_p = paddle velocity relative to water, m/s
(assume $v_p = 0.75 \, v$)
d = paddle wheel diameter, m
ω = paddle rotational speed = 3 rpm

$$d = 3.5 \text{ m} - 0.3 \text{ m} - 0.3 \text{ m} = 2.9 \text{ m}$$

$$v = \pi d \omega$$

$$= \left(\frac{\pi(2.9 \text{ m})}{\text{rev}}\right)(3 \text{ rpm})\left(\frac{1 \text{ min}}{60 \text{ s}}\right)$$

$$= 0.46 \text{ m/s}$$

$$v_p = \left(0.46 \; \frac{\text{m}}{\text{s}}\right)(0.75) = 0.345 \text{ m/s}$$

A = paddle area, m^2
C_d = drag coefficient, unitless = 1.8 for flat paddles
ρ = water density = 1000 kg/m^3

$$A = \frac{2P}{C_d \rho v_p^3}$$

$$A_1 = \frac{(2)\left(120\ \frac{\text{N·m}}{\text{s}}\right)\left(1\ \frac{\text{kg·m}}{\text{s}^2}\right)}{(1.8)\left(1000\ \frac{\text{kg}}{\text{m}^3}\right)\left(0.345\ \frac{\text{m}}{\text{s}}\right)^3 \text{N}} = \boxed{3.2\ \text{m}^2}$$

$$A_2 = \frac{(2)\left(30\ \frac{\text{N·m}}{\text{s}}\right)\left(1\ \frac{\text{kg·m}}{\text{s}^2}\right)}{(1.8)\left(1000\ \frac{\text{kg}}{\text{m}^3}\right)\left(0.345\ \frac{\text{m}}{\text{s}}\right)^3 \text{N}} = \boxed{0.81\ \text{m}^2}$$

The answer is D.

SOLUTION 3

3.1. A_s = settling zone surface area, m^2
Q = flow rate = $30\,000\ \text{m}^3/\text{d}$
q_o = overflow rate = $1.1\ \text{m}^3/\text{m}^2\text{·h}$

$$A_s = \frac{Q}{q_o}$$

$$= \frac{30\,000\ \frac{\text{m}^3}{\text{d}}}{\left(1.1\ \frac{\text{m}^3}{\text{m}^2\text{·h}}\right)\left(24\ \frac{\text{h}}{\text{d}}\right)}$$

$$= 1136\ \text{m}^2$$

$A_s = lw$ and since a $l{:}w$ of 4:1 is required, $l = 4w$. Also, $l \le 40$ m.

For one tank,

$$A_s = (4w)(w) = 4w^2 = 1136\ \text{m}^2$$
$$w = 16.9\ \text{m} \quad (17\ \text{m})$$
$$l = (4)(17\ \text{m}) = 68\ \text{m}$$

Since 68 m > 40 m, more than one tank is needed.

For two tanks,

$$A_s = 4w^2 = \frac{1136\ \text{m}^2}{2} = 568\ \text{m}^2/\text{tank}$$
$$w = 11.9\ \text{m} \quad (12\ \text{m})$$
$$l = (4)(12\ \text{m}) = 48\ \text{m}$$

Since 48 m > 40 m, more than two tanks are needed.

For three tanks,

$$A_s = 4w^2 = \frac{1136\ \text{m}^2}{3} = 379\ \text{m}^3/\text{tank}$$
$$w = 9.7\ \text{m} \quad (10\ \text{m})$$
$$l = (4)(10\ \text{m}) = 40\ \text{m}$$

This is OK. $\boxed{\text{Use three tanks.}}$

The answer is C.

3.2. From Prob. 3.1,

$$l = \boxed{40\ \text{m}}$$

$$w = \boxed{10\ \text{m}}$$

The answer is C.

3.3. Inlet and outlet zones, each with lengths equal to the settling zone depth of 2.5 m, are added to the settling zone length.

The sludge zone and freeboard are added to the settling zone depth. The sludge zone is typically selected at 0.5 m and the freeboard is 0.5 m.

The final total dimensions for each tank are

$$l = 40\ \text{m} + 2.5\ \text{m} + 2.5\ \text{m}$$
$$= \boxed{45\ \text{m}}$$
$$w = \boxed{10\ \text{m}}$$
$$d = 2.5\ \text{m} + 0.5\ \text{m} + 0.5\ \text{m}$$
$$= \boxed{3.5\ \text{m}}$$

The answer is B.

3.4. Q/tank = flow rate/tank = $10\,000\ \text{m}^3/\text{d}$
L_w = weir length, m
q_w = weir overflow rate, $14\ \text{m}^3/\text{m·h}$

$$L_w = \frac{\dfrac{Q}{\text{tank}}}{q_w}$$

$$= \frac{10\,000\ \frac{\text{m}^3}{\text{d}}}{\left(14\ \frac{\text{m}^3}{\text{m·h}}\right)\left(24\ \frac{\text{h}}{\text{d}}\right)}$$

$$= \boxed{29.8\ \text{m} \quad (30\ \text{m})}$$

The answer is B.

SOLUTION 4

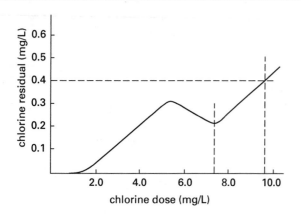

4.1. From the figure, breakpoint chlorination begins at a chlorine dose of $\boxed{7.5 \text{ mg/L.}}$

The answer is D.

4.2. From the figure, a free chlorine residual of 0.4 mg/L is obtained at a chlorine dose of $\boxed{9.6 \text{ mg/L.}}$

The answer is D.

SOLUTION 5

5.1.

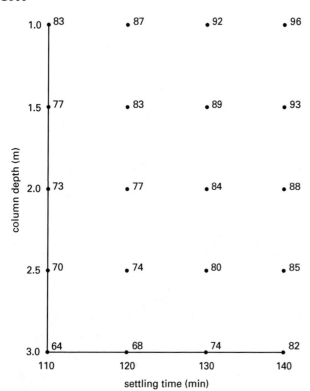

h = efficiency

h_o = incremental efficiency expressed as a fraction at t and Z_o

t = settling time (min)

Z_i = incremental depth (m) corresponding to an incremental change in efficiency (Δh) expressed as a fraction

Z_o = settling zone depth (m)

$$h = h_o + \frac{\Sigma(Z_i \Delta h)}{Z_o}$$

For trial 1, $t = 120$ min, $Z_o = 3.0$ m.

From the figure, at $t = 120$ min and $Z_o = 3.0$ m, $h_o = 68\%$ or 0.68.

Use the table below, developed from the figure, to find $\Sigma(Z_i \Delta h)$.

i	Z_i (m)	Δh (fraction)	$Z_i \Delta h$
1	2.5	0.06	0.15
2	2.0	0.03	0.06
3	1.5	0.06	0.09
4	1.0	0.04	0.04
			0.34

$$h = 0.68 + \frac{0.34}{3.0} = 0.793 \quad (79.3\%)$$

Since $79.3\% < 85\%$, repeat with increased t.

For trial 2, $t = 130$ min, $Z_o = 3.0$ m.

From the figure, at $t = 130$ min and $Z_o = 3.0$ m, $h_o = 74\%$ or 0.74.

Use the table below, developed from the figure, to find $\Sigma(Z_i \Delta h)$.

i	Z_i (m)	Δh (fraction)	$Z_i \Delta h$
1	2.5	0.06	0.15
2	2.0	0.04	0.08
3	1.5	0.05	0.08
4	1.0	0.03	0.03
			0.34

$$h = 0.74 + \frac{0.34}{3.0} = 0.853 \quad (85.3\%) \quad [\text{OK}]$$

$$t = \boxed{130 \text{ min}}$$

The answer is C.

5.2. V = settling zone volume, m³
Q = flow rate = 2000 m³/d

$$V = Qt$$

$$= \frac{\left(2000 \; \frac{m^3}{d}\right)(130 \text{ min})}{1440 \; \frac{\text{min}}{d}}$$

$$= \boxed{181 \text{ m}^3 \quad (180 \text{ m}^3)}$$

The answer is C.

5.3. A_s = settling zone surface area, m^2

$$A_s = \frac{V}{Z_o} = \frac{180 \text{ m}^3}{3.0 \text{ m}}$$

$$= \boxed{60 \text{ m}^2}$$

The answer is C.

SOLUTION 6

6.1. N_R = Reynolds number, unitless
v_s = filtering velocity = 4.6 m/h
ϕ = media shape factor, unitless = 0.82
d = media mean particle diameter
   = 0.00068 m
ρ = water density = 1000 kg/m^3
μ = water dynamic viscosity at 8°C
   = 1.39×10^{-3} kg/m·s

$$N_R = \frac{\phi d v_s \rho}{\mu}$$

$$= \frac{(0.82)(0.00068 \text{ m})\left(4.6 \; \frac{m}{h}\right)\left(1000 \; \frac{kg}{m^3}\right)}{\left(1.39 \times 10^{-3} \; \frac{kg}{m \cdot s}\right)\left(3600 \; \frac{s}{h}\right)}$$

$$= 0.51$$

f = friction factor, unitless
α = media packed bed porosity, unitless = 0.43

$$f = 150 \frac{(1 - \alpha)}{N_R} + 1.75$$

$$= \frac{(150)(1 - 0.43)}{0.51} + 1.75$$

$$= 169$$

h = clean filter head loss, m
L = media packed bed depth = 0.75 m
g = gravitational constant = 9.81 m/s^2

$$h = \frac{f(1 - \alpha)L \; v_s^2}{\phi \; \alpha^3 dg}$$

$$= \frac{(169)(1 - 0.43)(0.75 \text{ m})\left(4.6 \; \frac{m}{h}\right)^2 \left(\frac{1h}{3600 \text{ s}}\right)^2}{(0.82)(0.43)^3(0.00068 \text{ m})\left(9.81 \; \frac{m}{s^2}\right)}$$

$$= \boxed{0.27 \text{ m}}$$

The answer is B.

6.2. L_{fb} = media fluidized bed depth, m
α_{fb} = media fluidized bed porosity, unitless
   = 0.65

$$L_{fb} = \frac{L(1 - \alpha)}{1 - \alpha_{fb}}$$

$$= \frac{(0.75 \text{ m})(1 - 0.43)}{1 - 0.65}$$

$$= \boxed{1.2 \text{ m}}$$

The answer is B.

6.3. v_B = backwash velocity, m/s
v_t = media particle settling velocity
   = 0.10 m/s

$$v_B = v_t \left(1 - \frac{L(1 - \alpha)}{L_{fb}}\right)^{4.55}$$

$$= \left(0.10 \; \frac{m}{s}\right)\left(1 - \frac{(0.75 \text{ m})(1 - 0.43)}{1.22 \text{ m}}\right)^{4.55}$$

$$= 0.014 \text{ m/s}$$

v_B is usually expressed in units of m^3/m^2·min. Therefore, 0.014 m/s = $\boxed{0.84 \text{ m}^3/\text{m}^2 \cdot \text{min}}$

The answer is C.

6.4.
$$A_s = \frac{Q}{v_s}$$

$$= \frac{3000 \; \frac{m^3}{d}}{\left(4.6 \; \frac{m}{h}\right)\left(24 \; \frac{h}{d}\right)}$$

$$= \boxed{27 \text{ m}^2}$$

The answer is C.

SOLUTION 7

The annual flow rate is

$$\left(18\,000 \,\frac{m^3}{d}\right)\left(365 \,\frac{d}{yr}\right) = 6.57 \times 10^6 \text{ m}^3/\text{yr}$$

ion	concentration (mg/L)	mole weight (mg/mmol)	concentration (mmol/L)
Ca^{+2}	347	40	8.675
Mg^{+2}	129	24	5.375
HCO_3^-	1256	61	20.590

7.1. The following chemical equations are used for lime-soda ash softening.

$$Ca^{+2} + 2HCO_3^- + CaO \longrightarrow 2CaCO_3 \downarrow + H_2O \qquad [I]$$

$$Mg^{+2} + 2HCO_3^- + 2CaO \longrightarrow$$
$$2CaCO_3 \downarrow + Mg(OH)_2 \downarrow \qquad [II]$$

$$Mg^{+2} + \text{other ions} + CaO + H_2O \longrightarrow$$
$$Ca^{+2} + Mg(OH)_2 \downarrow + \text{other ions} \qquad [III]$$

$$Ca^{+2} + Na_2CO_3 \longrightarrow CaCO_3 \downarrow + 2Na^+ \qquad [IV]$$

From Eq. I, 8.675 mmol/L Ca^{+2} reacts with (2)(8.675 mmol/L) HCO_3^- and 8.675 mmol/L CaO to produce (2)(8.675 mmol/L) $CaCO_3$. Of the 20.59 mmol/L HCO_3^-, Eq. I consumes (2)(8.675 mmol/L) leaving 3.24 mmol/L for reaction in Eq. II.

From Eq. II, the remaining 3.24 mmol/L HCO_3^- reacts with ($\frac{1}{2}$)(3.24 mmol/L) Mg^{+2} and (3.24 mmol/L) CaO to produce (3.24 mmol/L) $CaCO_3$ and ($\frac{1}{2}$)(3.24 mmol/L) $Mg(OH)_2$. Of the 5.375 mmol/L Mg^{+2}, Eq. II consumes ($\frac{1}{2}$)(3.24 mmol/L) leaving 3.755 mmol/L for reaction in Eq. III.

From Eq. III, the remaining 3.755 mmol/L Mg^{+2} reacts with 3.755 mmol/L CaO to produce 3.755 mmol/L Ca^{+2} and 3.755 mmol/L $Mg(OH)_2$.

From Eq. IV, the 3.755 mmol/L Ca^{+2} produced in Eq. III reacts with 3.755 mmol/L Na_2CO_3 to produce 3.755 mmol/L $CaCO_3$.

The molecular weight of CaO is

$$40 \,\frac{mg}{mmol} + 16 \,\frac{mg}{mmol} = 56 \text{ mg/mmol}$$

The molecular weight of Na_2CO_3 is

$$(2)\left(23 \,\frac{mg}{mmol}\right) + 12 \,\frac{mg}{mmol}$$
$$+ (3)\left(16 \,\frac{mg}{mmol}\right) = 106 \text{ mg/mmol}$$

The cost of the CaO is

$$\frac{\begin{array}{c}\left(8.675 \,\dfrac{mmol}{L} + \left(\dfrac{1}{2}\right)\left(3.24 \,\dfrac{mmol}{L}\right) + 3.755 \,\dfrac{mmol}{L}\right) \\[2mm] \times \left(56 \,\dfrac{mg}{mmol}\right)\left(6.57 \times 10^6 \,\dfrac{m^3}{yr}\right)\left(\dfrac{\$82}{1000 \text{ kg}}\right)\end{array}}{\left(\dfrac{67\%}{100\%}\right)\left(\dfrac{1 \text{ m}^3}{1000 \text{ L}}\right)\left(10^6 \,\dfrac{mg}{kg}\right)}$$

$$= \$705{,}605/\text{yr}$$

The cost of the Na_2CO_3 is

$$\frac{\begin{array}{c}\left(3.755 \,\dfrac{mmol}{L}\right)\left(106 \,\dfrac{mg}{mmol}\right) \\[2mm] \times \left(6.57 \times 10^6 \,\dfrac{m^3}{yr}\right)\left(\dfrac{\$105}{1000 \text{ kg}}\right)\end{array}}{\left(\dfrac{98\%}{100\%}\right)\left(\dfrac{1 \text{ m}^3}{1000 \text{ L}}\right)\left(10^6 \,\dfrac{mg}{kg}\right)} = \$280{,}185/\text{yr}$$

The total cost of chemical reagents is

$$\frac{\$705{,}605}{\text{yr}} + \frac{\$280{,}185}{\text{yr}} = \boxed{\$985{,}790/\text{yr} \quad (\$990{,}000/\text{yr})}$$

The answer is D.

7.2. The molecular weight of $CaCO_3$ is

$$40 \,\frac{mg}{mmol} + 12 \,\frac{mg}{mmol} + (3)\left(16 \,\frac{mg}{mmol}\right)$$
$$= 100 \text{ mg/mmol}$$

The molecular weight of $Mg(OH)_2$ is

$$24 \,\frac{mg}{mmol} + (2)\left(16 \,\frac{mg}{mmol} + 1 \,\frac{mg}{mmol}\right) = 58 \text{ mg/mmol}$$

From Prob. 7.1, the total amount of $CaCO_3$ produced is

$$\frac{\begin{array}{c}\left((2)\left(8.675 \,\dfrac{mmol}{L}\right) + 3.24 \,\dfrac{mmol}{L} + 3.755 \,\dfrac{mmol}{L}\right) \\[2mm] \times \left(100 \,\dfrac{mg}{mmol}\right)\left(6.57 \times 10^6 \,\dfrac{m^3}{yr}\right)\end{array}}{\left(\dfrac{1 \text{ m}^3}{1000 \text{ L}}\right)\left(10^6 \,\dfrac{mg}{kg}\right)}$$

$$= 1.6 \times 10^7 \text{ kg/yr} \quad [\text{dry}]$$

From Prob. 7.1, the total amount of $Mg(OH)_2$ produced is

$$\frac{\left(\left(\frac{1}{2}\right)\left(3.24\ \dfrac{mmol}{L}\right)+3.755\ \dfrac{mmol}{L}\right)\left(58\ \dfrac{mg}{mmol}\right)\left(6.57\times10^6\ \dfrac{m^3}{yr}\right)}{\left(\dfrac{1\ m^3}{1000\ L}\right)\left(10^6\ \dfrac{mg}{kg}\right)}$$

$$= 2.0\times10^6\ kg/yr \quad [dry]$$

The total amount of sludge produced is

$$1.6\times10^7\ \frac{kg}{yr}\ [dry] + 2.0\times10^6\ \frac{kg}{yr}\ [dry]$$

$$= \boxed{1.8\times10^7\ kg/yr \quad [dry]}$$

The answer is C.

SOLUTION 8

8.1. Bypass enough water to allow 1.0 meq/L total hardness (TH) in the effluent.

influent TH = 4.5 meq/L + 2.0 meq/L = 6.5 meq/L

bypass flow rate = Q_B

treatment flow rate = 1000 m³/d - Q_B

The TH concentration of the water comprising the treatment flow rate = 0.0 meq/L.

$$1.0\ \frac{meq}{L} = \frac{Q_B\left(6.5\ \dfrac{meq}{L}\right) + \left(1000\ \dfrac{m^3}{d}-Q_B\right)\left(0.0\ \dfrac{meq}{L}\right)}{1000\ \dfrac{m^3}{d}}$$

$$Q_B = \boxed{154\ m^3/d \quad (150\ m^3/d)}$$

The answer is B.

8.2. $Q_{treatment} = 1000\ \dfrac{m^3}{d} - 154\ \dfrac{m^3}{d} = 846\ m^3/d$

The TH to be removed is

$$\left(6.5\ \frac{meq}{L}\right)\left(50\ \frac{mg}{meq}\right)\left(10^{-6}\ \frac{kg}{mg}\right)$$

$$\times\left(846\ \frac{m^3}{d}\right)\left(1000\ \frac{L}{m^3}\right) = 275\ kg\ TH/d$$

The resin volume is

$$\left(275\ \frac{kg\ TH}{d}\right)\left(\frac{1\ m^3}{60\ kg\ TH}\right) = \boxed{4.6\ m^3\ resin/d}$$

The answer is A.

8.3. Size and configure the system for a one-day operating cycle. One day requires 4.6 m³ of resin. The total tank surface area is

$$\frac{846\ \dfrac{m^3}{d}}{\left(0.4\ \dfrac{m^3}{m^2\cdot min}\right)\left(1440\ \dfrac{min}{d}\right)} = 1.47\ m^2$$

The tank diameter is 1.0 m; therefore, each tank's area is 0.785 m².

The total number of tanks is

$$\frac{1.47\ m^2}{0.785\ m^2} = \boxed{1.9\ tanks \quad (2\ tanks)}$$

The answer is B.

8.4. The daily media volume per tank is

$$\frac{4.6\ m^3}{2\ tanks} = 2.3\ m^3/tank$$

The bed depth per tank is

$$\frac{2.3\ m^3}{0.785\ m^2} = \boxed{2.9\ m}$$

The answer is C.

8.5. The regeneration salt required per tank is

$$\left(2.3\ \frac{m^3}{d}\right)\left(100\ \frac{kg\ NaCl}{m^3}\right) = 230\ kg\ NaCl/d$$

Using a 15% salt solution, the daily volume of solution is

$$\frac{230\ \dfrac{kg\ NaCl}{d}}{\left(\dfrac{15\ kg\ NaCl}{100\ kg\ water}\right)\left(1000\ \dfrac{kg}{m^3}\right)} = \boxed{1.5\ m^3/d}$$

The answer is B.

8.6. The regeneration cycle time in any single tank is

$$\frac{\left(1.5\ \dfrac{m^3}{d}\right)\left(1\ \dfrac{d}{cycle}\right)}{\left(0.04\ \dfrac{m^3}{m^2\cdot min}\right)(0.785\ m^2)} = \boxed{48\ min}$$

The answer is C.

SOLUTION 9

9.1. $\dfrac{X}{M} = \dfrac{\text{chemical mass removed, kg}}{\text{GAC mass used, kg}} = 5.37C_r^{0.58}$

$C_r = 0.005 \text{ mg/L}$

$\dfrac{X}{M} = (5.37)\left(0.005 \dfrac{\text{mg}}{\text{L}}\right)^{0.58}$

$\qquad = 0.25 \text{ kg chemical/kg GAC}$

$X = \text{chemical mass removed}$

$\quad = ((\Sigma \text{ influent chemical concentration, mg/L})$
$\qquad - 0.005 \text{ mg/L})Q$

$Q = \text{flow rate} = 2.0 \text{ m}^3/\text{min}$

$\Sigma \text{ influent chemical concentration} = 5.34 \text{ mg/L}$

$X = \left(5.34 \dfrac{\text{mg}}{\text{L}} - 0.005 \dfrac{\text{mg}}{\text{L}}\right)\left(2.0 \dfrac{\text{m}^3}{\text{min}}\right)$

$\qquad \times \left(1000 \dfrac{\text{L}}{\text{m}^3}\right)\left(1440 \dfrac{\text{min}}{\text{d}}\right)\left(10^{-6} \dfrac{\text{kg}}{\text{mg}}\right)$

$\qquad = 15.4 \text{ kg/d}$

$M = \text{GAC mass used} = \dfrac{15.4 \dfrac{\text{kg chemical}}{\text{d}}}{0.25 \dfrac{\text{kg chemical}}{\text{kg GAC}}}$

$\qquad = \boxed{61.6 \text{ kg GAC/d} \quad (62 \text{ kg GAC/d})}$

The answer is D.

9.2. The GAC change-out period is defined as from 60 to 90 d.

For 60 d, the required GAC mass is

$$\left(62 \dfrac{\text{kg GAC}}{\text{d}}\right)(60 \text{ d}) = 3720 \text{ kg GAC}$$

For 90 d, the required GAC mass is

$$\left(62 \dfrac{\text{kg GAC}}{\text{d}}\right)(90 \text{ d}) = 5580 \text{ kg GAC}$$

Select a standard GAC vessel with a capacity between 3720 kg GAC and 5580 kg GAC. The Calgon Model 7.5 has a GAC capacity of 4500 kg GAC. Therefore, select the Calgon Model 7.5 or equivalent.

GAC capacity $= \boxed{4500 \text{ kg GAC}}$

The answer is C.

9.3. The GAC change-out period is

$$\dfrac{4500 \text{ kg}}{62 \dfrac{\text{kg}}{\text{d}}} = \boxed{72.6 \text{ d} \quad (73 \text{ d})}$$

The answer is C.

9.4. Typical GAC bulk density is 450 kg/m^3. The empty bed contact time (EBCT) is

$$\dfrac{4500 \text{ kg}}{\left(450 \dfrac{\text{kg}}{\text{m}^3}\right)\left(2.0 \dfrac{\text{m}^3}{\text{min}}\right)} = \boxed{5 \text{ min}}$$

The answer is B.

SOLUTION 10

ion	concentration (mg/L)	mole weight (mg/mmol)	equivalence (meq/mmol)	concentration (meq/L)
Ca^{+2}	158	40	2	7.90
Mg^{+2}	47	24	2	3.92
Na^+	26	23	1	1.13
K^+	19	39	1	0.49
SO_4^{-2}	64	96	2	1.33
Cl^-	43	35.5	1	1.21
HCO_3^-	381	61	1	6.25

concentration (meq/L)
$= \dfrac{\text{concentration (mg/L)} \times \text{equivalence (meq/mmol)}}{\text{MW (mg/mmol)}}$

10.1.

$\Sigma \text{ cations} = 7.90 \dfrac{\text{meq}}{\text{L}} + 3.92 \dfrac{\text{meq}}{\text{L}}$

$\qquad\qquad + 1.13 \dfrac{\text{meq}}{\text{L}} + 0.49 \dfrac{\text{meq}}{\text{L}}$

$\qquad\quad = 13.44 \text{ meq/L}$

$\Sigma \text{ anions} = 1.33 \dfrac{\text{meq}}{\text{L}} + 1.21 \dfrac{\text{meq}}{\text{L}} + 6.25 \dfrac{\text{meq}}{\text{L}}$

$\qquad\quad = 8.79 \text{ meq/L}$

$\left(\dfrac{13.44 \dfrac{\text{meq}}{\text{L}} - 8.79 \dfrac{\text{meq}}{\text{L}}}{13.44 \dfrac{\text{meq}}{\text{L}}}\right)(100\%) = 35\%$

Since 35% is greater than 10%, the analysis is deficient in anions.

The answer is A.

10.2. The total hardness is

$$Ca^{+2} \frac{mg}{L} \text{ as } CaCO_3 + Mg^{+2} \frac{mg}{L} \text{ as } CaCO_3$$

$$= \left(7.90 \frac{meq}{L} + 3.92 \frac{meq}{L}\right)\left(\frac{50 \text{ mg as } CaCO_3}{1 \text{ meq}}\right)$$

$$= \boxed{591 \text{ mg/L as } CaCO_3 \quad (590 \text{ mg/L as } CaCO_3)}$$

The answer is C.

10.3. Carbonate hardness is the lesser of the concentration of total alkalinity and total hardness in mg/L as $CaCO_3$. All alkalinity for the water sample is from HCO_3^-.

The total alkalinity is

$$\left(6.25 \frac{meq}{L}\right)\left(\frac{50 \text{ mg as } CaCO_3}{1 \text{ meq}}\right)$$

$$= \boxed{312.5 \text{ mg/L as } CaCO_3 \quad (310 \text{ mg/L as } CaCO_3)}$$

310 mg/L as $CaCO_3$ is less than 590 mg/L as $CaCO_3$. Therefore, the carbonate hardness is equal to the total alkalinity.

The answer is B.

10.4. non-carbonate hardness

$$= \text{total hardness} - \text{carbonate hardness}$$

$$= 590 \frac{mg}{L} \text{ as } CaCO_3 - 310 \frac{mg}{L} \text{ as } CaCO_3$$

$$= \boxed{280 \text{ mg/L as } CaCO_3}$$

The answer is B.

SOLUTION 11

11.1. The molecular weight of HOCl is

$$1 \frac{g}{mol} + 16 \frac{g}{mol} + 35.5 \frac{g}{mol} = 52.5 \text{ g/mol}$$

$$\left(4 \frac{mg}{L}\right)\left(\frac{1 \text{ mol}}{52.5 \text{ g}}\right)\left(\frac{1 \text{ g}}{1000 \text{ mg}}\right) = 7.6 \times 10^{-5} \text{ mol/L}$$

$$HOCl \rightleftharpoons OCl^- + H^+$$

$$\frac{[H^+][OCl^-]}{[HOCl]} = 10^{-7.63}$$

$$[H^+] = [OCl^-] = x$$

$$[HOCl] = 7.6 \times 10^{-5} \frac{mol}{L} - x$$

$$\frac{x^2}{7.6 \times 10^{-5} \frac{mol}{L} - x} = 10^{-7.63} = 2.3 \times 10^{-8}$$

Solve for x using the quadratic formula.

$$x = 1.3 \times 10^{-6} \frac{mol}{L} = \boxed{1.3 \times 10^{-6} \text{ M}}$$

The answer is B.

11.2. From Prob. 11.1,

$$[OCl^-] = x = \boxed{1.3 \times 10^{-6} \text{ M}}$$

The answer is B.

11.3. $[HOCl] = 7.6 \times 10^{-5} \frac{mol}{L} - x$

$$= 7.6 \times 10^{-5} \frac{mol}{L} - 1.3 \times 10^{-6} \frac{mol}{L}$$

$$= 7.5 \times 10^{-5} \text{ mol/L}$$

Percent ionization of HOCl is

$$\left(1 - \frac{7.5 \times 10^{-5} \frac{mol}{L}}{7.6 \times 10^{-5} \frac{mol}{L}}\right)(100\%) = \boxed{1.3\%}$$

The answer is A.

SOLUTION 12

12.1.

ion	concentration (mg/L)	mole weight (g/mol)	valence (Z_i)	concentration (C_i, mol/L)	$C_i Z_i^2$
Ca^{+2}	46	40	2	0.00115	0.00460
Na^+	128	23	1	0.00557	0.00557
Cl^-	197	35.5	1	0.00555	0.00555
HCO_3^-	133	61	1	0.00218	0.00218
					0.01790

$$C_i \text{ (mol/L)} = \frac{\text{concentration (mg/L)}}{\text{MW (g/mol)} \times 1000 \frac{mg}{g}}$$

$$\text{ionic strength} = 0.5\Sigma C_i Z_i^2$$

$$= (0.5)(0.0179)$$

$$= \boxed{0.00895 \quad (0.0090)}$$

The answer is B.

12.2.
$$pCa^{+2} = -\log\left(Ca^{+2}\,\frac{mol}{L}\right)$$
$$= -\log(0.00115)$$
$$= 2.94$$
$$pAlk = -\log\left(HCO_3^-\,\frac{mol}{L}\right)$$
$$= -\log(0.00218)$$
$$= 2.66$$

SI = Langelier saturation index, unitless
$pK_2' - pK_s'$ = ionic strength and temperature constant
$$= 2.28 \text{ (given)}$$

$$SI = pH - [(pK_2' - pK_s') + pCa^{+2} + pAlk]$$
$$= 7.4 - 2.28 - 2.94 - 2.66$$
$$= \boxed{-0.48}$$

The answer is B.

12.3. The water is corrosive to iron pipes because SI is negative.

The answer is A.

SOLUTION 13

13.1. The population estimate from 1990 to 2030 is long-term. An appropriate method for long-term estimates where an historical record is used is the logistic-curve method.

Y_c = projected population for x years from the beginning historical record year (x_0).

$$Y_c = \frac{K}{1 + 10^{(a+bx)}}$$

K, a, and b are constants.

$$K = \frac{2Y_0Y_1Y_2 - Y_1^2(Y_0 + Y_2)}{Y_0Y_2 - Y_1^2}$$

Y_0, Y_1, Y_2 = populations from the historical record corresponding to years X_0, X_1, X_2 with the same time increment (n) between X_0 and X_1 and between X_1 and X_2.

$$K = \frac{\begin{array}{c}(2)(12\,200)(23\,500)(38\,600)\\ - (23\,500^2)(12\,200 + 38\,600)\end{array}}{(12\,200)(38\,600) - 23\,500^2} = 72\,803$$

$$a = \log\left(\frac{K - Y_0}{Y_0}\right)$$
$$= \log\left(\frac{72\,803 - 12\,200}{12\,200}\right)$$
$$= 0.696$$

$$b = \left(\frac{1}{n}\right)\left(\log\frac{Y_0(K - Y_1)}{Y_1(K - Y_0)}\right)$$
$$= \left(\frac{1}{20}\right)\left(\log\frac{(12\,200)(72\,803 - 23\,500)}{(23\,500)(72\,803 - 12\,200)}\right)$$
$$= -0.0187$$

2030 is 80 yr from the beginning of the historical record year 1950 (x_0). Therefore, $x = 80$ for Y_c calculation.

$$Y_c = \frac{72\,803}{1 + 10^{(0.696+(-0.0187)(80))}}$$
$$= \boxed{62{,}843 \text{ people} \quad (63{,}000 \text{ people})}$$

The answer is B.

13.2. Groundwater wells are low capital projects that can be easily constructed as demand increases. Therefore, a shorter design life ranging from 5 to 10 yr is appropriate.

The answer is A.

13.3. Water treatment plants are high capital projects, but can be expanded with relative ease if provisions for doing so are included in the original design. Building a plant to handle demand 40 yr in the future is hard to justify if expansion is planned for. Therefore, water treatment plants are designed to handle flows for 10 to 20 yr in the future.

The answer is B.

13.4. Pumping stations are easy to add as demand increases and, therefore, they are constructed to provide pumping capacity to meet projected demands for no more than about 10 yr in the future.

The answer is A.

SOLUTION 14

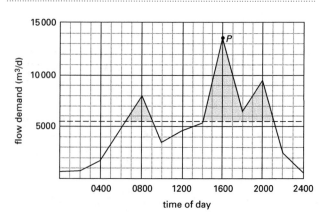

14.1. The pumping rate to meet peak demand occurs at point P on the figure.

$$\left(13\,500\ \frac{m^3}{d}\right)\left(\frac{1\ d}{86\,400\ s}\right)$$

$$= \boxed{0.156\ m^3/s \quad (0.16\ m^3/s)}$$

The answer is B.

14.2. The average pumping rate is

$$\left(3.6\ \frac{m^3}{min}\right)\left(1440\ \frac{min}{d}\right) = 5184\ m^3/d$$

The area under the curve but above the average pumping rate line on the figure (shaded area) is the required storage volume. Integrate the shaded area by counting squares. There are approximately 30 shaded squares in the figure.

$$\text{storage volume} = \left((30)\left(1000\ \frac{m^3}{d}\right)\right)(1\ h)\left(\frac{1\ d}{24\ h}\right)$$

$$= \boxed{1250\ m^3}$$

The answer is A.

SOLUTION 15

15.1. Solve using the Newton-Raphson method with the Hazen-Williams equation.

K = pipe constant, unitless
L = pipe length, m
C = Hazen-Williams constant = 100
D = pipe diameter, m

$$K = \frac{10.70L}{C^{1.85}D^{4.87}}$$

$$K_1 = \frac{(10.70)(1200\ m)}{(100)^{1.85}(0.305\ m)^{4.87}} = 832$$

$$K_2 = \frac{(10.70)(1800\ m)}{(100)^{1.85}(0.455\ m)^{4.87}} = 178$$

F = headloss function, unitless, at each iteration
h_A, h_{Bi} = head at nodes A and, for each iteration, at B, m
Q = flow rate = 0.3 m³/s
n = constant = 1.85

$$F_i = \left(\frac{h_A - h_{Bi}}{K_1}\right)^{1/n} + \left(\frac{h_A - h_{Bi}}{K_2}\right)^{1/n} - Q$$

$$= \left(\left(\frac{1}{832}\right)(100\ m - h_{Bi})\right)^{0.54}$$

$$+ \left(\left(\frac{1}{178}\right)(100\ m - h_{Bi})\right)^{0.54} - 0.3\ \frac{m^3}{s}$$

$$\left(\frac{\Delta F}{\Delta h}\right)_i = \frac{-(h_A - h_{Bi})^{(1/n)-1}}{n(K_1)^{1/n}} - \frac{(h_A - h_{Bi})^{(1/n)-1}}{n(K_2)^{1/n}}$$

$$= \frac{-(100\ m - h_{Bi})^{-0.46}}{(1.85)(832)^{0.54}} - \frac{(100\ m - h_{Bi})^{-0.46}}{(1.85)(178)^{0.54}}$$

$$h_{B(i+1)} = h_{Bi} - \frac{F_i}{\left(\frac{\Delta F}{\Delta h}\right)_i}$$

Iterations using the above equations are summarized in the table.

iteration, i	h_{Bi}(m)	F_i	$(\Delta F/\Delta h)_i$	$F_i/(\Delta F/\Delta h)_i$
1	80.00	0.1410	−0.0119	−11.850
2	91.85	−0.0286	−0.0180	1.589
3	90.26	−0.00118	−0.0166	0.0712
4	89.55	0.0104	−0.0161	−0.648
5	90.20	−0.0002	−0.0165	0.012

$$h_B = \boxed{90.20\ m \quad (90\ m)}$$

The answer is C.

15.2. $$Q = \left(\frac{\Delta h}{K}\right)^{1/n}$$

$$Q_1 = \left[(100\ m - 90\ m)\left(\frac{1}{832}\right)\right]^{0.54}$$

$$= \boxed{0.092\ m^3/s \quad (0.10\ m^3/s)}$$

$$Q_2 = \left[(100\ m - 90\ m)\left(\frac{1}{178}\right)\right]^{0.54}$$

$$= \boxed{0.21\ m^3/s \quad (0.20\ m^3/s)}$$

The flows in each pipe combine to create a total flow of 0.3 m³/s, which corresponds to the total initial flow through node A.

The answer is A.

SOLUTION 16

16.1. The maximum possible solute concentration will occur at the approximate water solubilities of benzene, toluene, and xylenes as shown in the following table.

chemical	water solubility (mg/L at 20°C)
benzene	1800
toluene	500
xylenes (average of the three isomers)	160

The answer is B.

16.2.
v_c = solute velocity, m/day
v_x = water velocity, m/day
r_f = retardation factor, unitless
$\quad = 1 + \rho_b K_d / \theta$
ρ_b = soil bulk density = 1.83 g/cm^3
θ = saturated soil porosity, unitless = 0.34
K_d = distribution coefficient, unitless
$\quad = K_{oc}$ TOC
K_{oc} = organic carbon partition coefficient, cm^3/g
TOC = soil total organic carbon
$\quad = 1.48 \times 10^{-4}$ g/g

$$v_c = v_x / r_f$$

As K_{oc} increases, v_c decreases. Chemicals with smaller K_{oc} values will move more quickly. K_{oc} values for benzene, toluene, and xylenes are summarized in the following table.

chemical	K_{oc} (cm^3/g)
benzene	97
toluene	242
xylenes (average of the three isomers)	501

With the smallest K_{oc}, benzene will move away from the source at the highest velocity.

The answer is A.

16.3.

$$v_c = \frac{v_x}{1 + \dfrac{\rho_b K_{oc} \text{TOC}}{\theta}}$$

$$= \frac{v_x}{1 + \dfrac{\left(1.83 \dfrac{\text{g}}{\text{cm}^3}\right)\left(97 \dfrac{\text{cm}^3}{\text{g}}\right)\left(1.48 \times 10^{-4} \dfrac{\text{g}}{\text{g}}\right)}{0.34}}$$

$$= \boxed{0.928\ v_x \quad (0.93\ v_x)}$$

The answer is C.

16.4.
$$v_c = \left(0.23 \dfrac{\text{m}}{\text{d}}\right)(0.93) = 0.21 \text{ m/d}$$

$$\frac{100 \text{ m}}{0.21 \dfrac{\text{m}}{\text{day}}} = \boxed{476 \text{ d} \quad (480 \text{ d})}$$

The answer is C.

SOLUTION 17

17.1.
v_c = solute velocity, m/d
v_x = water velocity = 0.37 m/d
r_f = retardation factor, unitless = 1.08

$$v_c = \frac{v_x}{r_f}$$

$$= \frac{0.37 \dfrac{\text{m}}{\text{d}}}{1.08} = \boxed{0.34 \text{ m/d}}$$

The answer is B.

17.2.
D_L = longitudinal hydrodynamic dispersion, m^2/d
α_L = dynamic dispersivity, m
$\quad$ (for flow paths less than 3500 m)
$\quad = 0.0175 L^{1.46}$
L = distance from the source to the point of interest
$\quad = 50$ m
v = average bulk groundwater or solute velocity, 0.34 m/d
D^* = effective diffusion, m^2/d

$$D_L = \alpha_L v + D^*$$

Assume D^* is negligible since the groundwater velocity indicates a relatively permeable soil.

$$D_L = \left(0.34 \dfrac{\text{m}}{\text{d}}\right)(0.0175)(50 \text{ m})^{1.46} = 1.80 \text{ m}^2/\text{d}$$

C = solute concentration at time t = 1.212 mg/L
C_0 = solute concentration at time zero
$\quad = 0.005$ mg/L (MCL for PCE)
t = travel time of interest, d
erfc = complimentary error function

$$C = \left(\frac{C_0}{2}\right) \text{erfc}\left(\frac{L - vt}{(2)(D_L t)^{1/2}}\right)$$

$$0.005 \frac{\text{mg}}{\text{L}} = \left(\frac{1.212 \dfrac{\text{mg}}{\text{L}}}{2}\right)$$

$$\times \text{erfc}\left(\frac{50 \text{ m} - \left(0.34 \dfrac{\text{m}}{\text{d}}\right) t}{(2)\left(\left(1.80 \dfrac{\text{m}^2}{\text{d}}\right) t\right)^{1/2}}\right)$$

$$0.00825 = \text{erfc}\left(\frac{50 \text{ m} - \left(0.34 \dfrac{\text{m}}{\text{d}}\right) t}{(2)\left(\left(1.80 \dfrac{\text{m}^2}{\text{d}}\right) t\right)^{1/2}}\right)$$

Let the erfc equal the following.

$$\left(\frac{50 \text{ m} - \left(0.34 \, \frac{\text{m}}{\text{d}}\right) t}{(2)\left(\left(1.80 \, \frac{\text{m}^2}{\text{d}}\right) t\right)^{1/2}} \right) = \text{erfc}(x) = 0.00825$$

From erfc reference tables, $x = 1.87$.

$$\frac{50 \text{ m} - \left(0.34 \, \frac{\text{m}}{\text{d}}\right) t}{(2)\left(\left(1.80 \, \frac{\text{m}^2}{\text{d}}\right) t\right)^{1/2}} = 1.87$$

Solve for t using the quadratic formula.

$$t = 47 \text{ d or } 465 \text{ d}$$

When compared to v_c, the reasonable answer is

$$t = \boxed{47 \text{ d}}$$

The answer is A.

Water Resources

PROBLEM 1

Calcium ion (Ca^{+2}) and carbonate ion (CO_3^{-2}) are present in a water sample at concentrations of 25 mg/L and 15 mg/L, respectively. The water temperature is 16°C.

1.1. What is the solubility product for $CaCO_3$ at 16°C?

(A) 5.8×10^{-9}

(B) 1.6×10^{-7}

(C) 2.5×10^{-4}

(D) 6.3×10^{-4}

1.2. What is the reaction quotient for calcium carbonate dissociation or precipitation under the above conditions?

(A) 5.8×10^{-9}

(B) 1.6×10^{-7}

(C) 2.5×10^{-4}

(D) 6.3×10^{-4}

1.3. Is calcium carbonate precipitating, dissociating, or at equilibrium under the above conditions?

(A) precipitating

(B) dissociating

(C) at equilibrium

(D) unable to determine from the information provided

PROBLEM 2

Analyses for BOD and dissolved oxygen are performed at several locations along a segment of river downstream from a municipal wastewater treatment plant outfall. The flow velocity in the river channel is 0.45 m/s and the average temperature of the river water is 15°C. River constants are $k_1 = 0.5$ d^{-1} and $k_2 = 0.9$ d^{-1} at 15°C. Typical data from representative monitoring stations along the river are summarized in the following table.

monitoring station	BOD_u (mg/L)	dissolved oxygen (mg/L)
A (upstream)	12	8.7
B (discharge)	18	7.9
C (downstream)	27	6.2
D (downstream)	31	5.8
E (downstream)	16	7.1

2.1. What is the oxygen deficit at station D?

(A) 2.9 mg/L

(B) 4.3 mg/L

(C) 5.8 mg/L

(D) 7.9 mg/L

2.2. What is the time of the critical oxygen sag point?

(A) 0.39 d

(B) 0.72 d

(C) 0.86 d

(D) 1.1 d

2.3. At what distance below the discharge does the critical oxygen sag point occur?

(A) 15 km

(B) 28 km

(C) 33 km

(D) 43 km

2.4. What is the oxygen deficit at the critical oxygen sag point?

(A) 3.2 mg/L

(B) 5.7 mg/L

(C) 7.6 mg/L

(D) 8.2 mg/L

PROBLEM 3

The dry weather average flow rate for a river is 8.7 m^3/s. During dry weather flow, the average COD concentration in the river is 32 mg/L. An industrial source continuously discharges 18 000 m^3/d of wastewater containing an average 342 mg/L COD concentration into the river.

3.1. What is the river flow rate after the industrial source discharge?

(A) 3.8 m³/s

(B) 8.9 m³/s

(C) 15 m³/s

(D) 21 m³/s

3.2. What is the COD concentration in the river after the industrial source discharge?

(A) 17 mg/L

(B) 39 mg/L

(C) 66 mg/L

(D) 92 mg/L

3.3. What is the COD mass loading in the river upstream of the industrial source discharge?

(A) 580 kg/d

(B) 6200 kg/d

(C) 24 000 kg/d

(D) 260 000 kg/d

3.4. What is the incremental COD mass loading to the river from the industrial source discharge?

(A) 580 kg/d

(B) 6200 kg/d

(C) 24 000 kg/d

(D) 260 000 kg/d

PROBLEM 4

A small natural lake occupies an area of 73 ha to an average depth of 30 m. The stream feeding the lake has an average flow of 0.21 m³/s with a total phosphorus concentration of 0.13 mg/L. The total phosphorus deposition rate in the lake is 10 m/year.

4.1. What is the annual total phosphorus loading to the lake?

(A) 0.050 kg/ha·m·yr

(B) 0.39 kg/ha·m·yr

(C) 12 kg/ha·m·yr

(D) 23 kg/ha·m·yr

4.2. What is the total phosphorus concentration in the lake?

(A) 0.062 mg/L

(B) 0.12 mg/L

(C) 3.6 mg/L

(D) 7.0 mg/L

4.3. What is the annual total phosphorus mass deposited in the lake sediments?

(A) 410 kg/yr

(B) 450 kg/yr

(C) 820 kg/yr

(D) 860 kg/yr

4.4. What is the annual total phosphorus mass lost from the lake in stream outflow?

(A) 410 kg/yr

(B) 450 kg/yr

(C) 820 kg/yr

(D) 860 kg/yr

PROBLEM 5

The profile of a river channel is shown in the figure. The channel flow was measured at distances from the east bank at 0.6 of the total depth in shallower sections and at 0.2 and 0.8 of the total depth in deeper sections. The flow measurement data are presented in the accompanying table.

distance from east bank

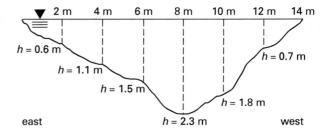

distance from east bank (m)	section depth, h (m)	flow velocity (m/s) at depth		
		$0.6h$	$0.2h$	$0.8h$
2	0.6	0.21	–	–
4	1.1	–	0.27	0.19
6	1.5	–	0.29	0.19
8	2.3	–	0.31	0.20
10	1.8	–	0.31	0.21
12	0.7	0.26	–	–

5.1. What is the average flow velocity in the river?

(A) 0.24 m/s

(B) 0.29 m/s

(C) 0.33 m/s

(D) 0.57 m/s

5.2. What is the average stream discharge?

(A) 4.0 m^3/s

(B) 4.8 m^3/s

(C) 5.5 m^3/s

(D) 9.5 m^3/s

PROBLEM 6

A water sample was collected from a stream with an average discharge of 43 500 m^3/d. The stream water sample was subjected to analysis for solids characterization and the resulting data are summarized below.

sample volume = 200 mL for TSS and VSS, 100 ml (filtered) for TDS

evaporation dish mass = 177.006 g

filter pad mass = 3.111 g

mass of evaporation dish + residue remaining after evaporation = 177.010 g

mass of filter pad + residue retained by filter (dried) = 3.142 g

mass of filter pad + residue retained by filter (combusted) = 3.139 g

6.1. What is the TSS concentration in the stream?

(A) 40 mg/L

(B) 140 mg/L

(C) 155 mg/L

(D) 195 mg/L

6.2. What is the VSS concentration in the stream?

(A) 15 mg/L

(B) 40 mg/L

(C) 140 mg/L

(D) 155 mg/L

6.3. What is the TDS concentration in the stream?

(A) 15 mg/L

(B) 20 mg/L

(C) 40 mg/L

(D) 75 mg/L

6.4. What is the total solids concentration in the stream?

(A) 195 mg/L

(B) 210 mg/L

(C) 295 mg/L

(D) 335 mg/L

6.5. What is the total solids loading in the stream?

(A) 8500 kg/d

(B) 9100 kg/d

(C) 13 000 kg/d

(D) 15 000 kg/d

6.6. What is the primary constituent of the solids loading to the stream?

(A) dissolved salts

(B) detritus and other organic matter

(C) inorganic sediment

(D) dissolved organic matter

PROBLEM 7

The following problems pertain to the management and regulation of wetlands in the United States.

7.1. What statute provided the initial regulatory authority for conservation of wetlands in the United States and remains the dominant statute today?

(A) Clean Water Act, Sec. 404

(B) Rivers & Harbors Appropriation Act, Sec. 10

(C) Federal Agricultural Improvement & Reform Act, Sec. 335

(D) North American Wetlands Conservation Act, all sections

7.2. Which federal agency has regulatory authority over wetlands in the United States?

(A) USEPA

(B) U.S. Army Corps of Engineers

(C) U.S. Department of Agriculture

(D) all of the above

7.3. What are the consequences of the 1985 Farm Bill "Swampbuster" provisions and 1996 amendments?

(A) Farmers are granted exemptions for wetlands under $^1/_4$ ac.

(B) Farmers can abandon converted cropland without restoring them to wetlands status.

(C) Agricultural subsidies can be withheld from farmers who violate "Swampbuster" rules.

(D) All exemptions for "good faith" violations are eliminated.

7.4. Which of the following are not included in the regulatory definition of wetlands?

(A) Areas that are inundated or saturated by surface or ground water.

(B) Areas that support vegetation typically adapted to life in saturated soil conditions.

(C) Areas that are frequently inundated and support a prevalence of vegetation adapted for life only in aerobic soil conditions.

(D) Wetlands generally include swamps, marshes, bogs, and similar areas.

PROBLEM 8

A water balance prepared for a natural wetland covering a 359 ha area is summarized in the following table. Berm height allows a maximum water depth of 80 cm in the wetland.

| | average estimated contribution | |
source	October–March	April–September
inputs		
direct precipitation	100 cm	55 cm
surface inflow	0.21 m^3/s	0.09 m^3/s
subsurface inflow	0.0043 m^3/s	–
outputs		
surface outflow	0.17 m^3/s	0.02 m^3/s
subsurface outflow	–	0.0073 m^3/s
evapotranspiration	40 cm	121 cm

8.1. What is the wetland minimum storage volume?

(A) 980 000 m^3

(B) 1 400 000 m^3

(C) 1 500 000 m^3

(D) 1 900 000 m^3

8.2. What is the annual change in storage volume between winter and summer months?

(A) 980 000 m^3

(B) 1 400 000 m^3

(C) 1 500 000 m^3

(D) 1 900 000 m^3

8.3. What is the maximum turnover rate during winter months?

(A) 0.32/mo

(B) 0.77/mo

(C) 3.1/mo

(D) 4.0/mo

8.4. What is the hydraulic retention time during summer months?

(A) 0.78 mo

(B) 3.1 mo

(C) 4.0 mo

(D) 8.0 mo

PROBLEM 9

A well constructed in a confined aquifer and screened through the entire aquifer thickness of 18 m was pumped at 0.75 m^3/min for 48 h. Time-drawdown observations at a well located 100 m away were recorded and the data plotted. Using a Theis type curve and the data plot, values for $h_o - h = 1.32$ m and $t = 47$ min were obtained.

9.1. What is the transmissivity of the aquifer?

(A) 3.3 m^2/d

(B) 6.5 m^2/d

(C) 33 m^2/d

(D) 65 m^2/d

9.2. What is the hydraulic conductivity of the aquifer?

(A) 0.18 m/d

(B) 0.36 m/d

(C) 1.8 m/d

(D) 3.6 m/d

9.3. What is the storativity of the aquifer?

(A) 4.3×10^{-5}

(B) 8.5×10^{-5}

(C) 4.3×10^{-4}

(D) 8.5×10^{-4}

PROBLEM 10

Groundwater monitoring wells have been constructed on a site as shown in the figure. The site's hydrogeologic setting is characterized by an unconfined aquifer with silty sand to a depth of 10 m. The average hydraulic conductivity for the aquifer is 0.42 m/d and the effective porosity is 0.34. Groundwater elevation data for the monitoring wells are provided in the accompanying table.

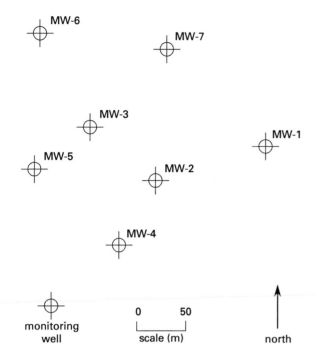

monitoring well

0   50

scale (m)

north

10.1. What is the direction of the groundwater gradient?

(A) N 45° E

(B) N 45° W

(C) S 45° E

(D) S 45° W

10.2. What is the slope of the groundwater gradient?

(A) 0.0058

(B) 0.0075

(C) 0.010

(D) 0.015

10.3. What is the actual groundwater flow velocity?

(A) 0.0072 m/d

(B) 0.0093 m/d

(C) 0.012 m/d

(D) 0.019 m/d

well	casing top elevation (m above mean sea level)	groundwater depth below casing top (m)
MW-1	49.77	4.74
MW-2	49.74	5.66
MW-3	49.59	5.59
MW-4	49.60	5.95
MW-5	49.09	5.57
MW-6	49.31	5.25
MW-7	49.63	4.62

Water Resources Solutions

SOLUTION 1

1.1. $CaCO_3 \longrightarrow Ca^{+2} + CO_3^{-2}$

The standard enthalpy (ΔH^o) for the reaction $= \Sigma \Delta H^o$ products $- \Sigma \Delta H^o$ reactants.

ΔH^o $Ca^{+2} = -543.0$ kJ/mol
ΔH^o $CO_3^{-2} = -676.3$ kJ/mol
ΔH^o $CaCO_3 = -1207.0$ kJ/mol

$$-543.0 \frac{\text{kJ}}{\text{mol}} + \left(-676.3 \frac{\text{kJ}}{\text{mol}}\right)$$
$$- \left(-1207.0 \frac{\text{kJ}}{\text{mol}}\right) = -12.3 \text{ kJ/mol}$$

$K_{T1} =$ solubility product at temperature 1
$\quad = 5.0 \times 10^{-9}$
$K_{T2} =$ solubility product at temperature 2
$\Delta H^o =$ standard enthalpy for the reaction
$\quad = -12.3$ kJ/mol
$T_1 =$ temperature 1 = 298K
$T_2 =$ temperature 2 = 289K
$R =$ universal gas constant = 8.314 J/mol K

$$\ln\left(\frac{K_{T2}}{K_{T1}}\right) = \frac{-\Delta H^o(T_1 - T_2)}{RT_1T_2}$$

$$\ln\left(\frac{K_{T2}}{5.0 \times 10^{-9}}\right) = \frac{\left(\begin{array}{c} -\left(-12.3 \dfrac{\text{kJ}}{\text{mol}}\right) \\ \times (298\text{K} - 289\text{K}) \\ \times \left(\dfrac{1000 \text{ J}}{1 \text{ kJ}}\right) \end{array}\right)}{\left(8.314 \dfrac{\text{J}}{\text{mol·K}}\right)(298\text{K})(289\text{K})}$$

$$= 0.15$$

$$K_{T2} = (5.0 \times 10^{-9})(e^{0.15})$$

$$= \boxed{5.8 \times 10^{-9}}$$

The answer is A.

1.2. $\quad Q =$ reaction quotient
$[Ca^{+2}] = Ca^{+2}$ concentration, mol/L
$[CO_3^{-2}] = CO_3^{-2}$ concentration, mol/L

$$Q = [Ca^{+2}][CO_3^{-2}]$$

$$[Ca^{+2}] = \frac{\left(25 \dfrac{\text{mg}}{\text{L}}\right)\left(\dfrac{1 \text{ g}}{1000 \text{ mg}}\right)}{40 \dfrac{\text{g}}{\text{mol}}} = 6.25 \times 10^{-4}$$

$$[CO_3^{-2}] = \frac{\left(15 \dfrac{\text{mg}}{\text{L}}\right)\left(\dfrac{1 \text{ g}}{1000 \text{ mg}}\right)}{12 \dfrac{\text{g}}{\text{mol}} + (3)\left(16 \dfrac{\text{g}}{\text{mol}}\right)} = 2.5 \times 10^{-4}$$

$$Q = (6.25 \times 10^{-4})(2.5 \times 10^{-4})$$

$$= \boxed{1.56 \times 10^{-7} \quad (1.6 \times 10^{-7})}$$

The answer is B.

1.3. K_{sp} is written for the reaction $CaCO_3 \rightarrow Ca^{+2} + CO_3^{-2}$. Since Q is greater than K_{sp}, the reaction is proceeding from right to left and $CaCO_3$ is precipitating.

The answer is A.

SOLUTION 2

2.1. Assume the water temperature along the river course is constant and the water salinity is low.

The saturated dissolved oxygen at 15°C and low salinity is 10.07 mg/L.

The oxygen deficit at station D is

$$10.07 \frac{\text{mg}}{\text{L}} - 5.8 \frac{\text{mg}}{\text{L}} = \boxed{4.27 \text{ mg/L} \quad (4.3 \text{ mg/L})}$$

The answer is B.

2.2. $\quad t_c =$ time of critical oxygen sag point, d
$D_o =$ dissolved oxygen deficit at discharge point
$L_o =$ BOD$_u$ at discharge point = 18 mg/L
$k_1 =$ river constant = 0.5 d^{-1}
$k_2 =$ river constant = 0.9 d^{-1}
$k_2 =$ river constant = 0.9 d^{-1}

$$D_o = 10.7 \frac{\text{mg}}{\text{L}} - 7.9 \frac{\text{mg}}{\text{L}} = 2.8 \text{ mg/L}$$

$$t_c = \left(\frac{1}{k_2 - k_1}\right) \ln\left[\left(\frac{k_2}{k_1}\right)\left(1 - \frac{D_o(k_2 - k_1)}{k_1 L_o}\right)\right]$$

$$= \left(\frac{1}{0.9 \text{ d}^{-1} - 0.5 \text{ d}^{-1}}\right) \ln\left[\left(\frac{0.9 \text{ d}^{-1}}{0.5 \text{ d}^{-1}}\right)\right.$$

$$\left. \times \left(1 - \frac{\left(2.8 \frac{\text{mg}}{\text{L}}\right)(0.9 \text{ d}^{-1} - 0.5 \text{ d}^{-1})}{(0.5 \text{ d}^{-1})\left(18 \frac{\text{mg}}{\text{L}}\right)}\right)\right]$$

$$= \boxed{1.14 \text{ d} \quad (1.1 \text{ d})}$$

The answer is D.

2.3. The distance below the discharge of critical oxygen sag point is

$$\left(0.45 \frac{\text{m}}{\text{s}}\right)(1.1 \text{ d})\left(86\,400 \frac{\text{s}}{\text{d}}\right)\left(\frac{1 \text{ km}}{1000 \text{ m}}\right)$$

$$= \boxed{42.8 \text{ km} \quad (43 \text{ km})}$$

The answer is D.

2.4. $a = -k_1 t_c = (-0.5 \text{ d}^{-1})(1.1 \text{ d}) = -0.55$
$b = -k_2 t_c = (-0.9 \text{ d}^{-1})(1.1 \text{ d}) = -0.99$
D_c = oxygen deficit at critical dissolved oxygen sag point

$$D_c = \frac{k_1 L_o(e^a - e^b)}{k_2 - k_1} + D_o e^b$$

$$= \frac{(0.5 \text{ d}^{-1})\left(18 \frac{\text{mg}}{\text{L}}\right)(e^{-0.55} - e^{-0.99})}{0.9 \text{ d}^{-1} - 0.5 \text{ d}^{-1}}$$

$$+ \left(2.8 \frac{\text{mg}}{\text{L}}\right)(e^{-0.99})$$

$$= \boxed{5.66 \text{ mg/L} \quad (5.7 \text{ mg/L})}$$

The answer is B.

SOLUTION 3

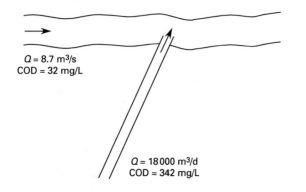

Q = 8.7 m³/s
COD = 32 mg/L

Q = 18 000 m³/d
COD = 342 mg/L

3.1. The total river flow after the discharge is

$$8.7 \frac{\text{m}^3}{\text{s}} + \left(18\,000 \frac{\text{m}^3}{\text{d}}\right)\left(\frac{1 \text{ d}}{86\,400 \text{ s}}\right) = \boxed{8.9 \text{ m}^3/\text{s}}$$

The answer is B.

3.2. The river COD concentration after the discharge is

$$\left(\frac{\left(8.7 \frac{\text{m}^3}{\text{s}}\right)\left(32 \frac{\text{mg}}{\text{L}}\right)}{+ \left(18\,000 \frac{\text{m}^3}{\text{d}}\right)\left(342 \frac{\text{mg}}{\text{L}}\right)\left(\frac{1 \text{ d}}{86\,400 \text{ s}}\right)}\right) \Big/ 8.9 \frac{\text{m}^3}{\text{s}}$$

$$= \boxed{39.3 \text{ mg/L} \quad (39 \text{ mg/L})}$$

The answer is B.

3.3. The river COD mass loading upstream of the discharge is

$$\left(8.7 \frac{\text{m}^3}{\text{s}}\right)\left(32 \frac{\text{mg}}{\text{L}}\right)\left(1000 \frac{\text{L}}{\text{m}^3}\right)$$

$$\times \left(10^{-6} \frac{\text{kg}}{\text{mg}}\right)\left(86\,400 \frac{\text{s}}{\text{d}}\right)$$

$$= \boxed{24\,054 \text{ kg/d} \quad (24\,000 \text{ kg/d})}$$

The answer is C.

3.4. Incremental COD mass loading to the river from the discharge is

$$\left(18\,000 \frac{\text{m}^3}{\text{d}}\right)\left(342 \frac{\text{mg}}{\text{L}}\right)\left(1000 \frac{\text{L}}{\text{m}^3}\right)\left(10^{-6} \frac{\text{kg}}{\text{mg}}\right)$$

$$= \boxed{6156 \text{ kg/d} \quad (6200 \text{ kg/d})}$$

The answer is B.

SOLUTION 4

4.1. The annual total phosphorus loading is

$$\left(\frac{\left(0.21 \frac{\text{m}^3}{\text{s}}\right)\left(0.13 \frac{\text{mg}}{\text{L}}\right)\left(1000 \frac{\text{L}}{\text{m}^3}\right)}{\times \left(10^{-6} \frac{\text{kg}}{\text{mg}}\right)\left(86\,400 \frac{\text{s}}{\text{d}}\right)}\right) \Big/ (73 \text{ ha})(30 \text{ m})\left(\frac{1 \text{ yr}}{365 \text{ d}}\right)$$

$$= \boxed{0.39 \text{ kg/ha·m·yr}}$$

The answer is B.

4.2. Q = inlet and outlet flow rate
(assume they are equal)
$= 0.21 \text{ m}^3/\text{s}$
C_{oTP} = inflow total phosphorus concentration
$= 0.13 \text{ mg/L}$
v_s = total phosphorus deposition rate
$= 10 \text{ m/yr}$
A_s = lake surface area $= 73 \text{ ha}$
C_P = total phosphorus concentration in
the lake

$$C_P = \frac{QC_{oTP}}{Q + v_s A_s}$$

$$= \frac{\left(0.21 \, \dfrac{\text{m}^3}{\text{s}}\right)\left(0.13 \, \dfrac{\text{mg}}{\text{L}}\right)}{\left(\begin{array}{c} 0.21 \, \dfrac{\text{m}^3}{\text{s}} + \left(10 \, \dfrac{\text{m}}{\text{yr}}\right)(73 \text{ ha})\left(\dfrac{1 \text{ yr}}{365 \text{ d}}\right) \\ \times \left(\dfrac{1 \text{ d}}{86\,400 \text{ s}}\right)\left(10\,000 \, \dfrac{\text{m}^2}{\text{ha}}\right) \end{array}\right)}$$

$$= \boxed{0.062 \text{ mg/L}}$$

The answer is A.

4.3. The annual total phosphorus mass deposited to sediments is

$$\frac{\left(0.21 \, \dfrac{\text{m}^3}{\text{s}}\right)\left(0.13 \, \dfrac{\text{mg}}{\text{L}}\right) - \left(0.21 \, \dfrac{\text{m}^3}{\text{s}}\right)\left(0.062 \, \dfrac{\text{mg}}{\text{L}}\right)}{\left(\dfrac{1 \text{ m}^3}{1000 \text{ L}}\right)\left(10^6 \, \dfrac{\text{mg}}{\text{kg}}\right)\left(\dfrac{1 \text{ d}}{86\,400 \text{ s}}\right)\left(\dfrac{1 \text{ yr}}{365 \text{ d}}\right)}$$

$$= \boxed{450 \text{ kg/yr}}$$

The answer is B.

4.4. The annual total phosphorus mass lost from the lake in stream outflow is

$$\frac{\left(0.21 \, \dfrac{\text{m}^3}{\text{s}}\right)\left(0.062 \, \dfrac{\text{mg}}{\text{L}}\right)}{\left(\dfrac{1 \text{ m}^3}{1000 \text{ L}}\right)\left(10^6 \, \dfrac{\text{mg}}{\text{kg}}\right)\left(\dfrac{1 \text{ d}}{86\,400 \text{ s}}\right)\left(\dfrac{1 \text{ yr}}{365 \text{ d}}\right)}$$

$$= \boxed{410 \text{ kg/yr}}$$

The answer is A.

SOLUTION 5

5.1.

distance from east bank (m)	section depth, h (m)	flow velocity (m/s) at depth			total section velocity (m/s)
		0.6h	0.2h	0.8h	
2	0.6	0.21	–	–	0.21
4	1.1	–	0.27	0.19	0.23
6	1.5	–	0.29	0.19	0.24
8	2.3	–	0.31	0.20	0.26
10	1.8	–	0.31	0.21	0.26
12	0.7	0.26	–	–	0.26
					1.46

For shallower sections (depth less than 1.0 m), the total section velocity is equal to the velocity at 0.6h.

For deeper sections (depth greater than 1.0 m), the total section velocity is

$$\frac{\text{velocity at } 0.2h + \text{velocity at } 0.8h}{2}$$

The average flow velocity in the river is

$$\frac{1.46}{6} = \boxed{0.24 \text{ m/s}}$$

The answer is A.

5.2.

distance from east bank (m)	total section velocity (m/s)	section area (m²)	section flow rate (m³/s)
2	0.21	2 m × 0.6 m	0.252
4	0.23	2 m × 1.1 m	0.506
6	0.24	2 m × 1.5 m	0.720
8	0.26	2 m × 2.3 m	1.196
10	0.26	2 m × 1.8 m	0.936
12	0.26	2 m × 0.7 m	0.364
			3.97

The average stream discharge is

$$\boxed{3.97 \text{ m}^3/\text{s} \quad (4.0 \text{ m}^3/\text{s})}$$

The answer is A.

SOLUTION 6

6.1.

$$\text{TSS} = \frac{(3.142 \text{ g} - 3.111 \text{ g})\left(1000 \, \dfrac{\text{mg}}{\text{g}}\right)\left(1000 \, \dfrac{\text{mL}}{\text{L}}\right)}{200 \text{ mL}}$$

$$= \boxed{155 \text{ mg/L}}$$

The answer is C.

6.2.

$$\text{VSS} = \frac{(3.142 \text{ g} - 3.139 \text{ g})\left(1000 \frac{\text{mg}}{\text{g}}\right)\left(1000 \frac{\text{mL}}{\text{L}}\right)}{200 \text{ mL}}$$

$$= \boxed{15 \text{ mg/L}}$$

The answer is A.

6.3.

$$\text{TDS} = \frac{\left(\begin{array}{c}(177.010 \text{ g} - 177.006 \text{ g}) \\ \times \left(1000 \frac{\text{mg}}{\text{g}}\right)\left(1000 \frac{\text{mL}}{\text{L}}\right)\end{array}\right)}{100 \text{ mL}}$$

$$= \boxed{40 \text{ mg/L}}$$

The answer is C.

6.4. The total solids concentration is

$$\text{TSS} + \text{TDS} = 155 \frac{\text{mg}}{\text{L}} + 40 \frac{\text{mg}}{\text{L}}$$

$$= \boxed{195 \text{ mg/L}}$$

The answer is A.

6.5. The total solids loading is

$$\left(195 \frac{\text{mg}}{\text{L}}\right)\left(43\,500 \frac{\text{m}^3}{\text{d}}\right)\left(1000 \frac{\text{L}}{\text{m}^3}\right)\left(10^{-6} \frac{\text{kg}}{\text{mg}}\right)$$

$$= \boxed{8483 \text{ kg/d} \quad (8500 \text{ kg/d})}$$

The answer is A.

6.6. Comparing the concentrations of TSS and VSS, a relatively small portion of the TSS is volatile, so most of the TSS is composed of inorganic sediments. Also, the concentration of TSS is nearly four times greater than that of TDS. Therefore, the primary constituent of solids loading to the stream is inorganic sediment.

The answer is C.

SOLUTION 7

7.1. Although other federal statutes provide regulatory authority for wetlands conservation in the United States, the initial statute and the one that remains dominant today is the Clean Water Act, Sec. 404.

The answer is A.

7.2. The USEPA, the US Army Corps of Engineers, and the US Department of Agriculture all have regulatory authority over wetlands in the United States.

The answer is D.

7.3. The 1985 Farm Bill "Swampbuster" provisions and the 1996 amendments provide authority to withhold agricultural subsidies from farmers who violate the rules.

The answer is C.

7.4. Wetlands are defined as areas that are inundated or saturated by surface or ground water and that support vegetation typically adapted to life in saturated soil conditions. Wetlands generally include swamps, marshes, bogs, and similar areas.

The definition does not include areas that are frequently inundated and support a prevalence of vegetation adapted for life only in aerobic soil conditions.

The answer is C.

SOLUTION 8

8.1. The change in volume is

$$\Delta V = \text{inputs} - \text{outputs}$$

The change in volume from October to March is

$$\frac{(100 \text{ cm} - 40 \text{ cm})(359 \text{ ha})\left(10\,000 \frac{\text{m}^2}{\text{ha}}\right)\left(\frac{1 \text{ m}}{100 \text{ cm}}\right)}{6 \text{ mo}}$$

$$+ \left(0.21 \frac{\text{m}^3}{\text{s}} + 0.0043 \frac{\text{m}^3}{\text{s}} - 0.17 \frac{\text{m}^3}{\text{s}}\right)$$

$$\times \left(86\,400 \frac{\text{s}}{\text{d}}\right)\left(30 \frac{\text{d}}{\text{mo}}\right)$$

$$= 473\,826 \text{ m}^3/\text{mo} \quad \text{[increase]}$$

The change in volume from April to September is

$$\frac{(55 \text{ cm} - 121 \text{ cm})(359 \text{ ha})\left(10\,000 \frac{\text{m}^2}{\text{ha}}\right)\left(\frac{1 \text{ m}}{100 \text{ cm}}\right)}{6 \text{ mo}}$$

$$+ \left(0.09 \frac{\text{m}^3}{\text{s}} - 0.02 \frac{\text{m}^3}{\text{s}} - 0.0073 \frac{\text{m}^3}{\text{s}}\right)$$

$$\times \left(86\,400 \frac{\text{s}}{\text{d}}\right)\left(30 \frac{\text{d}}{\text{mo}}\right)$$

$$= -232\,382 \text{ m}^3/\text{mo} \quad \text{[decrease]}$$

From October to March, the wetland gains water at a faster rate than it loses water from April to September. Therefore, at some point during the October to March period the wetland will be at its maximum depth of 80 cm, corresponding to its maximum storage volume.

The maximum storage volume is

$$(80 \text{ cm})(359 \text{ ha}) \left(10\,000 \ \frac{\text{m}^2}{\text{ha}}\right) \left(\frac{1 \text{ m}}{100 \text{ cm}}\right)$$
$$= 2\,872\,000 \text{ m}^3 \quad (2\,900\,000 \text{ m}^3)$$

The minimum storage volume is

$$2\,900\,000 \text{ m}^3 - \left(232\,382 \ \frac{\text{m}^3}{\text{mo}}\right) (6 \text{ mo})$$
$$= \boxed{1\,505\,708 \text{ m}^3 \quad (1\,500\,000 \text{ m}^3)}$$

The answer is C.

8.2. The annual storage volume changes from a maximum of $2\,872\,000$ m^3 to $1\,500\,000$ m^3.

The annual volume change is

$$2\,900\,000 \text{ m}^3 - 1\,500\,000 \text{ m}^3$$
$$= \boxed{1\,400\,000 \text{ m}^3}$$

The answer is B.

8.3. To estimate the maximum turnover rate, assume the minimum volume occurs from October to March.

$$\text{turnover} = \frac{\text{input}}{\text{volume}}$$

$$= \frac{(100 \text{ cm})(359 \text{ ha}) \left(10\,000 \ \frac{\text{m}^2}{\text{ha}}\right) \left(\frac{1 \text{ m}}{100 \text{ cm}}\right)}{(6 \text{ mo})(1\,500\,000 \text{ m}^3)}$$
$$+ \frac{\left(\left(0.21 \ \frac{\text{m}^3}{\text{s}} + 0.0043 \ \frac{\text{m}^3}{\text{s}}\right) \times \left(86\,400 \ \frac{\text{s}}{\text{d}}\right) \left(30 \ \frac{\text{d}}{\text{mo}}\right)\right)}{(1\,500\,000 \text{ m}^3)}$$
$$= \boxed{0.77/\text{mo}}$$

The answer is B.

8.4. $\qquad \text{hydraulic retention time} = \dfrac{\text{volume}}{\text{inputs}}$

The hydraulic retention time (summer) is

$$\frac{(1\,500\,000 \text{ m}^3)}{\left(\begin{array}{c} \left(\frac{55 \text{ cm}}{6 \text{ mo}}\right) (359 \text{ ha}) \left(10\,000 \ \frac{\text{m}^2}{\text{ha}}\right) \left(\frac{1 \text{ m}}{100 \text{ cm}}\right) \\ + \left(0.09 \ \frac{\text{m}^3}{\text{s}}\right) \left(86\,400 \ \frac{\text{s}}{\text{d}}\right) \left(30 \ \frac{\text{d}}{\text{mo}}\right) \end{array}\right)}$$
$$= \boxed{4.0 \text{ mo}}$$

The answer is C.

SOLUTION 9

9.1. $\quad T = $ transmissivity, m^2/d
$\quad Q = $ pumping rate $= 0.75$ m^3/min
$\quad W(u) = $ well function for u, dimensionless, assume $= 1$ for $W(u)$ and u
$\quad h_o - h = $ water table drawdown $= 1.32$ m

$$T = \frac{QW(u)}{4\pi(h_o - h)}$$
$$= \frac{\left(0.75 \ \frac{\text{m}^3}{\text{min}}\right) (1) \left(1440 \ \frac{\text{min}}{\text{d}}\right)}{4\pi(1.32 \text{ m})}$$
$$= \boxed{65 \text{ m}^2/\text{d}}$$

The answer is D.

9.2. $K = $ hydraulic conductivity, m/d
$\quad b = $ aquifer thickness $= 18$ m

$$K = \frac{T}{b} = \frac{65 \ \frac{\text{m}^2}{\text{d}}}{18 \text{ m}} = \boxed{3.6 \text{ m/d}}$$

The answer is D.

9.3. $S = $ storativity, unitless
$\quad t = $ time since pumping began (from Theis type curve)
$\quad \quad = 47$ min

$$S = \frac{4Tut}{r^2}$$
$$= \frac{(4) \left(65 \ \frac{\text{m}^2}{\text{d}}\right) (1)(47 \text{ min})}{(100 \text{ m})^2 \left(1440 \ \frac{\text{min}}{\text{d}}\right)}$$
$$= \boxed{8.5 \times 10^{-4}}$$

The answer is D.

SOLUTION 10

well	casing top elevation (m above mean sea level)	groundwater depth below casing top (m)	groundwater elevation (m above mean sea level)
MW-1	49.77	4.74	45.03
MW-2	49.74	5.66	44.08
MW-3	49.59	5.59	44.00
MW-4	49.60	5.95	43.65
MW-5	49.09	5.57	43.52
MW-6	49.31	5.25	44.06
MW-7	49.63	4.62	45.01

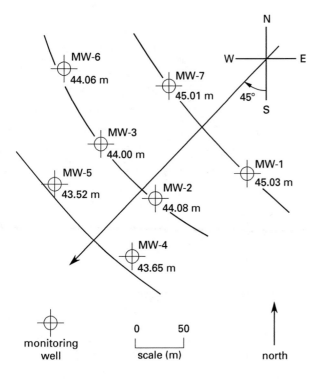

10.1. From the figure, the direction of the groundwater gradient is S 45° W.

The answer is D.

10.2. From the figure, the gradient, i, is

$$\frac{\Delta\text{elevation}}{\Delta\text{distance}} = \frac{45.0 \text{ m} - 43.5 \text{ m}}{150 \text{ m}} = \boxed{0.010}$$

The answer is C.

10.3. v_x = actual groundwater velocity, m/d
K = hydraulic conductivity = 0.42 m/d
n_e = effective porosity = 0.34

$$v_x = \frac{Ki}{n_e}$$

$$= \frac{\left(0.42 \ \dfrac{\text{m}}{\text{d}}\right)(0.010)}{0.34}$$

$$= \boxed{0.012 \text{ m/d}}$$

The answer is C.

Section II
Air

- Ambient Air

- Emissions Sources

- Control Strategies

Ambient Air

The following problems address criteria pollutants as defined and regulated under the Clean Air Act.

1.1. What are the criteria pollutants?

(A) ground level ozone, carbon dioxide, sulfur dioxide, small particulates, nitrogen dioxide, and radon

(B) ground level ozone, carbon monoxide, sulfur trioxide, small particulates, nitrogen dioxide, and radon

(C) ground level ozone, carbon monoxide, sulfur trioxide, small particulates, nitrogen dioxide, and lead

(D) ground level ozone, carbon monoxide, sulfur dioxide, small particulates, nitrogen dioxide, and lead

1.2. What is the primary source of nitrogen dioxide as a pollutant in the atmosphere?

(A) Nitrogen dioxide results from fuel release and thermal formation of nitric oxide that is subsequently oxidized to nitrogen dioxide after being emitted to the atmosphere.

(B) Nitrogen dioxide is released directly to the atmosphere from fuel release and thermal formation during combustion.

(C) Nitrogen dioxide results from thermal formation of ammonia during oxygen-deficient combustion where the ammonia is oxidized to nitrogen dioxide after being emitted to the atmosphere.

(D) Nitrogen dioxide is created from atmospheric nitrogen gas during combustion of fossil fuels when excess oxygen is present at temperatures between 400°C and 600°C.

1.3. Which of the following are considered significant precursors to photochemical oxidants such as ozone?

(A) volatile organic compounds (VOCs) and sulfur oxides

(B) volatile organic compounds (VOCs) and nitrogen oxides

(C) volatile organic compounds (VOCs) and carbon oxides

(D) volatile organic compounds (VOCs) and sulfur and nitrogen oxides

1.4. What characteristic of particulate matter presents the greatest potential human health hazard?

(A) particle size, especially where diameters exceed 10 μm

(B) potential for adsorption of toxic organics onto particle surfaces

(C) persistence which allows their re-release into the atmosphere

(D) origin from uncontrollable phenomena such as wind-borne erosion

The air pollutants shown in the following table represent the maximum concentrations measured in a metropolitan area of the United States during a single day.

pollutant	duration (h)	concentration
O_3	1	0.21 ppm
NO_2	1	0.08 ppm
CO	8	9.5 ppm
SO_2	24	0.10 ppm
PM-10	24	100 μg/m^3

2.1. Which pollutants exceed the national ambient air quality standards (NAAQS)?

(A) ozone and particulate

(B) ozone and carbon monoxide

(C) carbon monoxide and particulate

(D) sulfur dioxide

2.2. Which pollutant has the highest pollutant standard index (PSI) subindex value?

(A) ozone

(B) carbon monoxide

(C) sulfur dioxide

(D) particulate

2.3. What is the pollutant standard index (PSI) value for the day?

- (A) 75
- (B) 121
- (C) 205
- (D) 484

2.4. What is the descriptor corresponding to the pollutant standard index (PSI) value for the day?

- (A) moderate
- (B) unhealthful
- (C) very unhealthful
- (D) hazardous

PROBLEM 3

The following questions pertain to pollutant definitions and emission control actions under the Clean Air Act.

3.1. What do PM-2.5 and PM-10 describe?

- (A) particulates measured over 2.5 h and 10 h intervals
- (B) particulate matter with diameters smaller than 2.5 μm and 10 μm
- (C) total pollutant mass composited from 2.5 L and 10 L mylar sample bags
- (D) preventive maintenance and monitoring of new pollution control equipment at 2.5 months and 10 months after being placed in service

3.2. What emission controls must be applied by new sources in nonattainment areas?

- (A) best available control technology (BACT)
- (B) advanced pollutant control and management (APCAM)
- (C) lowest achievable emission rate (LAER)
- (D) comprehensive health and environment abatement technology (CHEAT)

3.3. What distinguishes "national ambient air quality standards" (NAAQS) from "new source performance standards" (NSPS)?

- (A) NAAQS define acceptable concentrations of pollutants in the air, whereas NSPS define allowable rates at which pollutants can be emitted.
- (B) NSPS define acceptable concentrations of pollutants in the air, whereas NAAQS define allowable rates at which pollutants can be emitted.
- (C) NAAQS define acceptable concentrations in ambient air outside of specified urban areas, whereas NSPS define air quality goals within specified urban areas.
- (D) NSPS apply exclusively to limit emissions from mobile sources, whereas NAAQS are not source specific and define acceptable concentrations of pollutants in the air overall.

3.4. How do primary and secondary pollutants differ from primary and secondary standards?

- (A) Primary and secondary standards define emission limits for primary and secondary pollutants, respectively.
- (B) Primary and secondary standards are regulatory emission limits, while primary and secondary pollutants are defined by their source or by reactions in the atmosphere.
- (C) Primary and secondary pollutants are defined by their health and environmental impacts, while primary and secondary standards limit emissions based on technological and economic factors.
- (D) Primary and secondary standards are regulatory emission limits, while primary and secondary pollutants are defined based on technological factors required for their control.

PROBLEM 4

A series of three samplers operated over a 24 hr period are described by the data presented in the following table.

parameter	sampler 1	sampler 2	sampler 3
particle size collected, μm	unrestricted	<10	<2.5
clean filter mass, g	9.87	10.03	9.96
filter mass after 24 h, g	9.93	10.05	9.97
initial air flow, m³/min	1.0	1.0	1.0
final air flow, m³/min	0.87	0.91	0.90

4.1. What is the respirable particulate concentration in the sampled air?

- (A) 7.3 μg/m³
- (B) 15 μg/m³
- (C) 23 μg/m³
- (D) 45 μg/m³

4.2. What is the PM-10 concentration in the sampled air?

- (A) 7.3 μg/m³
- (B) 15 μg/m³
- (C) 23 μg/m³
- (D) 45 μg/m³

4.3. What is the total particulate concentration in the sampled air?

(A) $7.3 \ \mu g/m^3$

(B) $15 \ \mu g/m^3$

(C) $23 \ \mu g/m^3$

(D) $45 \ \mu g/m^3$

PROBLEM 5

A typical wind rose is shown in the figure.

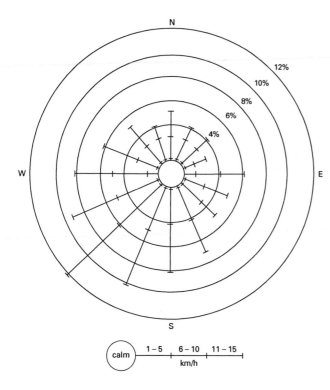

5.1. From which direction does the wind most frequently blow?

(A) northeast

(B) east-northeast

(C) southwest

(D) west-southwest

5.2. What is the maximum velocity for winds originating from the south-southeast?

(A) 1 to 5 km/h

(B) 6 to 10 km/h

(C) 11 to 15 km/h

(D) greater than 15 km/h

5.3. What percent of the time does wind blow to the west?

(A) 2%

(B) 4%

(C) 6%

(D) 8%

5.4. What is the most common velocity for winds blowing from the north-northeast?

(A) calm

(B) 1 to 5 km/h

(C) 6 to 10 km/h

(D) 11 to 15 km/h

PROBLEM 6

The diagrams presented in the figure show common atmospheric stability patterns. The dashed line represents the dry adiabatic lapse rate and the solid line represents the ambient lapse rate.

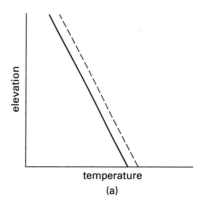

(a)

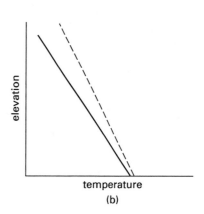

(b)

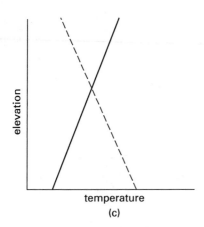

(c)

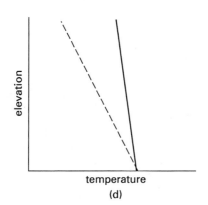

(d)

6.1. Which diagram illustrates a neutral atmospheric stability?

(A) a

(B) b

(C) c

(D) d

6.2. Which diagram illustrates a subadiabatic atmospheric stability?

(A) a

(B) b

(C) c

(D) d

6.3. Which diagram illustrates an inversion?

(A) a

(B) b

(C) c

(D) d

6.4. Which diagram illustrates conditions when a looping plume is observed?

(A) a

(B) b

(C) c

(D) d

6.5. Which diagram illustrates the most stable atmospheric conditions?

(A) a

(B) b

(C) c

(D) d

PROBLEM 7

Particulate matter with an average diameter of 10 μm is present in 23°C ambient air. The particle density is 500 kg/m^3, and the air pressure is 1 atm.

7.1. What is the settling velocity of the particulate matter?

(A) 5.9×10^{-6} m/s

(B) 1.5×10^{-3} m/s

(C) 2.9×10^{-3} m/s

(D) 1.7×10^{-2} m/s

7.2. What is the Reynolds number associated with the settling particle?

(A) 1.2×10^{-13}

(B) 9.4×10^{-4}

(C) 1.2×10^{-3}

(D) 120

7.3. Does Stokes' law apply?

(A) No, neither settling velocity nor particle diameter criteria are satisfied.

(B) No, settling velocity criteria are not satisfied.

(C) No, particle diameter criteria are not satisfied.

(D) Yes, both settling velocity and particle diameter criteria are satisfied.

PROBLEM 8

Air component concentrations are presented in the table. The ambient temperature and pressure are 30°C and 0.98 atm, respectively.

component	concentration (ppmv)
N_2	780 900
O_2	209 400
Ar	9350
CO_2	350
	1 000 000

8.1. What is the apparent molecular weight of the air?

(A) 15 g/mol

(B) 29 g/mol

(C) 36 g/mol

(D) 140 g/mol

8.2. What is the molar volume of the air?

(A) 2.7×10^{-7} m^3/mol

(B) 2.5×10^{-3} m^3/mol

(C) 0.025 m^3/mol

(D) 250 m^3/mol

8.3. What is the density of the air?

(A) 580 g/m^3

(B) 1100 g/m^3

(C) 1400 g/m^3

(D) 5600 g/m^3

8.4. What is the specific volume of the air?

(A) 0.00018 m^3/g

(B) 0.00071 m^3/g

(C) 0.00091 m^3/g

(D) 0.0017 m^3/g

8.5. What is the solubility of the air in water?

(A) 11 mg/L

(B) 21 mg/L

(C) 26 mg/L

(D) 100 mg/L

PROBLEM 9

Indoor radon exposure presents one of the more significant cancer risks to humans. Radon progeny are electrically charged and attach to airborne particulates that are inhaled and deposited in the lungs.

9.1. What is the primary source of radon gas in most buildings?

(A) masonry building materials

(B) soil surrounding foundations and crawl spaces

(C) concrete walls and slabs

(D) lead containing paints and sealers

9.2. At what concentration in the indoor air does the USEPA recommend immediate action to mitigate radon exposure?

(A) 4 pCi/L

(B) 20 pCi/L

(C) 100 pCi/L

(D) 200 pCi/L

9.3. What is the dominant mechanism for radon transport into buildings?

(A) It is carried with individuals as they move from one location to another.

(B) It is the density differential between radon gas and air.

(C) It is the presence of forced air ventilation systems.

(D) It is the pressure differential between inside and outside air.

9.4. What radioactive particles are associated with radon?

(A) alpha particles

(B) beta particles

(C) gamma particles

(D) delta particles

PROBLEM 10

Asbestos and formaldehyde are commonly associated with indoor air pollution.

10.1. What is the primary source of asbestos and formaldehyde in indoor air?

(A) movement of individuals in to and out of buildings

(B) materials used in building construction

(C) ventilation systems with outside air intakes

(D) furniture and office supplies

10.2. What are the symptoms of low-level indoor formaldehyde exposure?

(A) lung cancer

(B) cancer of the mucosa of the eyes, mouth, and nose

(C) headaches, sinus congestion, depression

(D) hyperactivity, increased appetite, hair loss

10.3. Which federal statute addresses asbestos exposure in schools?

(A) RCRA—Resource Conservation & Recovery Act

(B) OSHA—Occupational Safety & Health Act

(C) AHERA—Asbestos Hazard Emergency Response Act

(D) ACBMA—Asbestos Contaminated Building Mitigation Act

10.4. When does asbestos present a health risk?

(A) in any form when accessible for human contact

(B) in friable form

(C) in nonfriable form

(D) in unprocessed mineral form only

10.5. What are the effects of nonoccupational asbestos exposure?

(A) lung cancer and mesothelioma

(B) cancer of the mucosa of the eyes, mouth, and nose

(C) asbestosis

(D) headache, sinus congestion, depression

Ambient Air Solutions

SOLUTION 1

1.1. The criteria pollutants are ground level ozone, carbon monoxide, sulfur dioxide, small particulates, nitrogen dioxide, and lead.

The answer is D.

1.2. The primary source of nitrogen dioxide as a pollutant is fuel release and thermal formation of nitric oxide that is subsequently oxidized to nitrogen dioxide after being emitted to the atmosphere.

The answer is A.

1.3. Volatile organic compounds (VOCs) and nitrogen oxides are considered significant precursors to photochemical oxidants such as ozone.

The answer is B.

1.4. The characteristic of particulate matter that presents the greatest potential human health hazard is the potential for adsorption of toxic organics onto particle surfaces.

The answer is B.

SOLUTION 2

The following table presents the NAAQS and pollutant standard index (PSI) values for the measured pollutants.

pollutant	NAAQS concentration	measured concentration	subindex
O_3	0.08 ppm	0.21 ppm	205
NO_2	–	0.08 ppm	–
CO	9 ppm	9.5 ppm	108
SO_2	0.14 ppm	0.10 ppm	82
PM-10	150 $\mu g/m^3$	100 $\mu g/m^3$	75

$$\text{subindex } O_3 = 200 + (300 - 200)\left(\frac{0.21 - 0.20}{0.40 - 0.20}\right)$$
$$= 205$$
$$\text{subindex } CO = 100 + (200 - 100)\left(\frac{9.5 - 9}{15 - 9}\right)$$
$$= 108$$

$$\text{subindex } SO_2 = 50 + (100 - 50)\left(\frac{0.10 - 0.03}{0.14 - 0.03}\right)$$
$$= 82$$
$$\text{subindex PM-10} = 50 + (100 - 50)\left(\frac{100 - 50}{150 - 50}\right)$$
$$= 75$$

2.1. From the table, ozone and carbon monoxide exceed the NAAQS.

The answer is B.

2.2. From the table, ozone has the highest PSI subindex value at 205.

The answer is A.

2.3. The PSI for the day is equal to the highest subindex value. From the table, the highest subindex is for ozone at 205. Therefore, the PSI for the day is 205.

The answer is C.

2.4. The PSI is 205, so the descriptor would be "very unhealthful."

The answer is C.

SOLUTION 3

3.1. PM-2.5 and PM-10 describe particulate matter with diameters smaller than 2.5 μm and 10 μm.

The answer is B.

3.2. The lowest achievable emission rate (LAER) must be applied to control emissions from new sources in nonattainment areas.

The answer is C.

3.3. NAAQS define acceptable concentrations of pollutants in the air, whereas NSPS define allowable rates at which pollutants can be emitted.

The answer is A.

3.4. Primary and secondary standards are regulatory emission limits, while primary and secondary pollutants are defined by their source or by reactions in the atmosphere.

The answer is B.

SOLUTION 4

4.1. Respirable particles are less than 2.5 μm. Use data from sampler 3.

The average air flow is

$$(0.5)\left(1.0\ \frac{\text{m}^3}{\text{min}} + 0.90\ \frac{\text{m}^3}{\text{min}}\right) = 0.95\ \text{m}^3/\text{min}$$

The concentration is

$$\frac{(9.97\ \text{g} - 9.96\ \text{g})\left(10^6\ \frac{\mu\text{g}}{\text{g}}\right)}{\left(0.95\ \frac{\text{m}^3}{\text{min}}\right)(24\ \text{h})\left(60\ \frac{\text{min}}{\text{h}}\right)} = \boxed{7.3\ \mu\text{g}/\text{m}^3}$$

The answer is A.

4.2. PM-10 particles are less than 10 μm. Use data from sampler 2.

The average air flow is

$$(0.5)\left(1.0\ \frac{\text{m}^3}{\text{min}} + 0.91\ \frac{\text{m}^3}{\text{min}}\right) = 0.955\ \text{m}^3/\text{min}$$

The concentration is

$$\frac{(10.05\ \text{g} - 10.03\ \text{g})\left(10^6\ \frac{\mu\text{g}}{\text{g}}\right)}{\left(0.955\ \frac{\text{m}^3}{\text{min}}\right)(24\ \text{h})\left(60\ \frac{\text{min}}{\text{h}}\right)}$$

$$= \boxed{14.5\ \mu\text{g}/\text{m}^3 \quad (15\ \mu\text{g}/\text{m}^3)}$$

The answer is B.

4.3. Total particulate is measured by sampler 1.

The average air flow is

$$(0.5)\left(1.0\ \frac{\text{m}^3}{\text{min}} + 0.87\ \frac{\text{m}^3}{\text{min}}\right) = 0.935\ \text{m}^3/\text{min}$$

The concentration is

$$\frac{(9.93\ \text{g} - 9.87\ \text{g})\left(10^6\ \frac{\mu\text{g}}{\text{g}}\right)}{\left(0.935\ \frac{\text{m}^3}{\text{min}}\right)(24\ \text{h})\left(60\ \frac{\text{min}}{\text{h}}\right)}$$

$$= \boxed{44.6\ \mu\text{g}/\text{m}^3 \quad (45\ \mu\text{g}/\text{m}^3)}$$

The answer is D.

SOLUTION 5

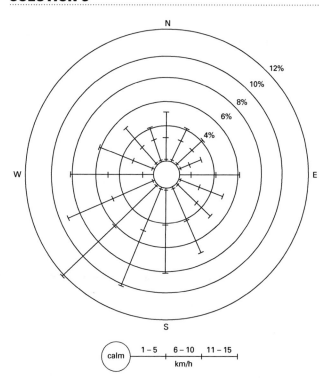

5.1. From the figure, the wind most frequently blows from the southwest.

The answer is C.

5.2. From the figure, the maximum velocity for winds originating from the south-southeast is between 11 and 15 km/h.

The answer is C.

5.3. From the figure, the wind blows to the west 6% of the time.

The answer is C.

5.4. From the figure, the most common velocity for winds blowing from the north-northeast is 6 to 10 km/h.

The answer is C.

SOLUTION 6

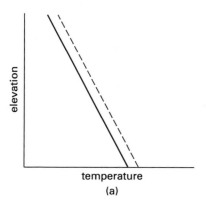

(a)

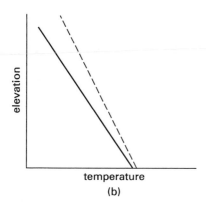

(b)

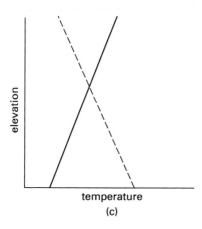

(c)

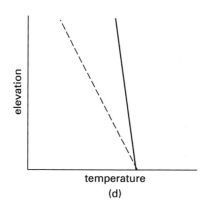

(d)

6.1. Diagram (a) shows the ambient lapse rate to be the same as the dry adiabatic lapse rate. This illustrates neutral atmospheric stability.

The answer is A.

6.2. Diagram (d) shows the ambient lapse rate to be greater than the dry adiabatic lapse rate. This illustrates subadiabatic atmospheric stability.

The answer is D.

6.3. Diagram (c) shows the ambient lapse rate with a positive slope and the dry adiabatic lapse rate with a negative slope. This illustrates an inversion.

The answer is C.

6.4. A looping plume occurs under superadiabatic conditions illustrated by diagram (b).

The answer is B.

6.5. The most stable atmospheric conditions occur under an inversion as illustrated by diagram (c).

The answer is C.

SOLUTION 7

7.1. v_s = particle settling velocity, m/s
   g = gravitational constant = 9.81 m/s^2
   ρ_p = particle density = 500 kg/m^3
   ρ_a = air density
       = 1.195 kg/m^3 at 23°C and 1 atm
   d_p = particle diameter = 10 μm
   μ = air viscosity = 1.84×10^{-5} kg/m·s

$$v_s = \frac{g(\rho_p - \rho_a)d_p^2}{18\mu}$$

$$= \frac{\left(9.81 \ \frac{m}{s^2}\right)\left(500 \ \frac{kg}{m^3} - 1.195 \ \frac{kg}{m^3}\right)(10 \times 10^{-6} \ m)^2}{(18)\left(1.84 \times 10^{-5} \ \frac{kg}{m \cdot s}\right)}$$

$$= \boxed{1.5 \times 10^{-3} \ m/s}$$

The answer is B.

7.2.

$$\text{Re} = \text{Reynolds number}$$

$$= \frac{v_s d_p \rho_a}{\mu}$$

$$= \frac{\left(1.6 \times 10^{-3} \, \frac{\text{m}}{\text{s}}\right) (10 \times 10^{-6} \, \text{m}) \left(1.3 \, \frac{\text{kg}}{\text{m}^3}\right)}{1.7 \times 10^{-5} \, \frac{\text{kg}}{\text{m·s}}}$$

$$= \boxed{1.2 \times 10^{-3}}$$

The answer is C.

7.3. Stokes' law applies for settling velocities less than 1.0 m/s and particle diameters between 0.1 and 100 μm. For this problem, the settling velocity is 1.6×10^{-3} m/s and the particle diameter is 10 μm. Therefore, Stokes' law applies.

The answer is D.

SOLUTION 8

8.1.

component	volumetric fraction	component mole weight (g/mol)	fractional mole weight (g/mol)
N_2	0.7809	28	21.87
O_2	0.2094	32	6.70
Ar	0.00935	40	0.37
CO_2	0.00035	44	0.015
			28.96

The fractional molecular weight is

$$\text{(volumetric fraction)}$$
$$\times \text{(component mole weight, g/mol)}$$

The apparent molecular weight is

$$\boxed{28.96 \text{ g/mol} \quad (29 \text{ g/mol})}$$

The answer is B.

8.2. Assume air behaves as an ideal gas.

$P = \text{pressure} = 0.98$ atm
$V = \text{volume, m}^3$
$n = \text{number of moles}$
$R = \text{gas law constant} = 8.2 \times 10^{-5} \text{ m}^3\text{·atm/mol·K}$
$T = \text{temperature} = (30°\text{C} + 273)\text{K} = 303\text{K}$

$$PV = nRT$$

The molar volume is

$$\frac{V}{n} = \frac{RT}{P}$$

$$= \frac{\left(8.2 \times 10^{-5} \, \frac{\text{m}^3\text{·atm}}{\text{mol·K}}\right) (303\text{K})}{0.98 \text{ atm}}$$

$$= \boxed{0.025 \text{ m}^3/\text{mol}}$$

The answer is C.

8.3. $\rho = \text{density, g/m}^3$
$\quad\;\; MW = \text{molecular weight} = 29$ g/mol

$$\rho = \frac{P(MW)}{RT}$$

$$= \frac{(0.98 \text{ atm}) \left(29 \, \frac{\text{g}}{\text{mol}}\right)}{\left(8.2 \times 10^{-5} \, \frac{\text{m}^3\text{·atm}}{\text{mol·K}}\right) (303\text{K})}$$

$$= \boxed{1144 \text{ g/m}^3 \quad (1100 \text{ g/m}^3)}$$

The answer is B.

8.4. $V_s = \text{specific volume, m}^3/\text{g}$

$$V_s = \frac{1}{\rho} = \frac{1}{1100 \, \frac{\text{g}}{\text{m}^3}} = \boxed{0.00091 \text{ m}^3/\text{g}}$$

The answer is C.

8.5. Assume Henry's law applies.

$x = \text{mole fraction}$
$P_g = \text{partial pressure} = 0.98$ atm
$K_H = \text{Henry's constant}$
$\quad\;\; = \dfrac{7.71 \times 10^4 \text{ atm}}{\text{mol fraction}}$ at 30°C

$$x = \frac{P_g}{K_H} = \frac{0.98 \text{ atm}}{\left(\dfrac{7.71 \times 10^4 \text{ atm}}{\text{mol fraction}}\right)}$$

$$= 1.3 \times 10^{-5} \text{ mol fraction}$$

$$x = \frac{\text{mol air}}{\text{mol air} + \text{mol water}}$$

$$= 1.3 \times 10^{-5}$$

Assume a 1 L sample of water with water density = 1000 g/L.

The molecular weight of water is

$$(2) \left(1 \, \frac{\text{g}}{\text{mol}}\right) + 16 \, \frac{\text{g}}{\text{mol}} = 18 \text{ g/mol}$$

$$\text{mol water} = \frac{\left(1000 \ \frac{g}{L}\right)(1 \ L)}{18 \ \frac{g}{\text{mol}}} = 55.6 \ \text{mol}$$

$$1.3 \times 10^{-5} = \frac{\text{mol air}}{\text{mol air} + 55.6 \ \text{mol}}$$

$$\text{mol air} = 7.2 \times 10^{-4}$$

The solubility of air in water is

$$\frac{\left(29 \ \frac{g}{\text{mol}}\right)(7.2 \times 10^{-4} \ \text{mol})\left(1000 \ \frac{mg}{g}\right)}{1 \ L}$$

$$= \boxed{20.9 \ \text{mg/L} \quad (21 \ \text{mg/L})}$$

The answer is B.

SOLUTION 9

9.1. The primary source of radon gas in most buildings is soil surrounding foundations and crawlspaces.

The answer is B.

9.2. The USEPA recommends immediate action to mitigate radon exposure when concentrations exceed 4 pCi/L.

The answer is A.

9.3. The dominant mechanism for radon transport into buildings is the pressure differential between inside and outside air.

The answer is D.

9.4. The radioactive particles associated with radon are alpha particles.

The answer is A.

SOLUTION 10

10.1. Building materials are the primary sources of asbestos and formaldehyde in indoor air.

The answer is B.

10.2. Symptoms of low-level formaldehyde exposure include headaches, sinus congestion, and depression.

The answer is C.

10.3. Congress passed the Asbestos Hazard Emergency Response Act (AHERA) in 1986 specifically to address asbestos in schools.

The answer is C.

10.4. Asbestos presents a health risk when it is in friable form allowing it to potentially become airborne.

The answer is B.

10.5. Nonoccupational exposure to asbestos can cause lung cancer and mesothelioma. Asbestosis is typically associated with occupational exposure.

The answer is A.

Emissions Sources

PROBLEM 1

The following questions relate to general air pollution source issues.

1.1. What are the six contaminants that typically constitute the major components of urban air pollution?

(A) smog, HCl, NO_x, SO_x, Pb, and O_3

(B) smog, NO_x, SO_x, CO_2, Ar, and particulate

(C) NO_x, SO_x, CO_x, Ar, O_3, and particulate

(D) NO_x, SO_x, CO, Pb, O_3, and particulate

1.2. Which of the following does not contribute significantly to pollution from combustion?

(A) incomplete combustion

(B) combustion in air

(C) moisture in fuels and water vapor in air

(D) compounds other than C and H in fuels

1.3. Which of the following is most typically associated with automobiles?

(A) photochemical smog

(B) industrial smog

(C) carbonaceous smog

(D) sulfurous smog

1.4. Which represents the largest source of air pollution in urban and other industrialized regions of the world?

(A) dust from wind and erosion

(B) plant and animal bioeffluents

(C) fossil fuel combustion

(D) vapors from architectural coatings, gasoline, cleaning fluids, and other solvent-based compounds

PROBLEM 2

Smoke plumes may be qualitatively and semiquantitatively assessed by certified observers.

2.1. How is the opacity of black or gray smoke reported?

(A) Ringlemann number

(B) pararosaniline number

(C) percent opacity

(D) percent polarization

2.2. How is opacity of white smoke reported?

(A) Ringlemann number

(B) pararosaniline number

(C) percent opacity

(D) percent polarization

2.3. With what accuracy are opacity observations reported?

(A) 2.5% opacity or 1/8 Ringlemann number

(B) 5% opacity or 1/4 Ringlemann number

(C) 7.5% opacity or 3/8 Ringlemann number

(D) 10% opacity or 1/2 Ringlemann number

PROBLEM 3

A 75 m high stack emits a plume at 20°C. The ground temperature is 16°C, and the prevailing lapse rate is -10.1°C/km up to an altitude of 250 m. Above 250 m, the prevailing lapse rate is 19.8°C/km. Assume the gas exits the stack at near-zero velocity, and assume a negligible wind speed.

3.1. What type of atmospheric stability conditions exit?

(A) subadiabatic

(B) inversion

(C) inversion over superadiabatic

(D) superadiabatic

3.2. How high will the plume rise?

(A) 250 m

(B) 390 m

(C) 410 m

(D) 730 m

3.3. What type of plume will be formed?

(A) coning

(B) fanning

(C) fumigation

(D) looping

3.4. What is the maximum mixing depth?

(A) 250 m

(B) 390 m

(C) 410 m

(D) 730 m

3.5. How is the dry adiabatic lapse rate affected by a water-saturated atmosphere?

(A) The dry adiabatic lapse rate will increase in a water-saturated atmosphere.

(B) The dry adiabatic lapse rate will decrease in a water-saturated atmosphere.

(C) The dry adiabatic lapse rate is not directly influenced by atmospheric moisture.

(D) The dry adiabatic lapse rate changes in a water-saturated atmosphere and that change may be an increase or a decrease.

PROBLEM 4

Stack emissions from a manufacturing facility contain SO_2 at 0.18 ppm V/V and 95°C. The ambient air temperature is 18°C, and the air pressure is 1 atm.

4.1. What is the SO_2 concentration in $\mu g/m^3$ at the stack?

(A) 15 $\mu g/m^3$

(B) 190 $\mu g/m^3$

(C) 380 $\mu g/m^3$

(D) 540 $\mu g/m^3$

4.2. If dispersion is neglected, what is the SO_2 concentration in $\mu g/m^3$ in the air at ambient temperature?

(A) 480 $\mu g/m^3$

(B) 540 $\mu g/m^3$

(C) 1200 $\mu g/m^3$

(D) 2000 $\mu g/m^3$

4.3. If it is raining, what is the likely fate of the SO_2?

(A) The SO_2 will remain as a gas.

(B) The SO_2 will form sulfuric acid with the rainwater.

(C) The SO_2 will form sulfurous acid with the rainwater.

(D) The SO_2 will condense from a gas to a liquid.

PROBLEM 5

Sulfur trioxide (SO_3) exists in the atmosphere at 458 $g/10^6$ m^3 of air at the beginning of a rain storm. The natural rainwater pH is 5.5, and the ambient temperature and pressure are 25°C and 1 atm, respectively. Assume the rainwater is 100% efficient in scrubbing the SO_3 from the air and that each cubic meter of air is scrubbed by 0.27 m^3 of rainwater over the duration of the storm.

5.1. What will be the sulfate concentration in the rainwater at the end of the storm?

(A) 1.4 $\mu g/L$

(B) 2.0 $\mu g/L$

(C) 21 $\mu g/L$

(D) 30 $\mu g/L$

5.2. What will be the pH of the rainwater at the end of the storm?

(A) 1.4

(B) 1.7

(C) 5.5

(D) 6.9

PROBLEM 6

A power plant burns coal with 3.8% sulfur at 11 tons/h. The combustion products are emitted through a stack with an effective height of 175 m. The wind speed at 10 m above the ground is 8 m/s, and atmospheric conditions are neutral to slightly unstable.

6.1. What is the SO_2 emission rate?

(A) 0.21 kg/s

(B) 0.53 kg/s

(C) 3.7 kg/s

(D) 5.6 kg/s

6.2. What is the location of the maximum ground level SO_2 concentration?

(A) 2600 m

(B) 3700 m

(C) 15 000 m

(D) 22 000 m

6.3. What is the maximum ground level SO_2 concentration?

(A) 92 $\mu g/m^3$

(B) 760 $\mu g/m^3$

(C) 8600 $\mu g/m^3$

(D) 130 000 $\mu g/m^3$

6.4. What is the ground level SO_2 concentration 1.5 km downwind of the stack along the plume centerline?

(A) 2.7 $\mu g/m^3$

(B) 78 $\mu g/m^3$

(C) 36 000 $\mu g/m^3$

(D) 50 000 $\mu g/m^3$

6.5. What is the ground level SO_2 concentration 1.5 km downwind of the stack at a crosswind distance of 200 m?

(A) 42 $\mu g/m^3$

(B) 520 $\mu g/m^3$

(C) 780 $\mu g/m^3$

(D) 19 000 $\mu g/m^3$

PROBLEM 7

A remedial action program for contaminated soils has identified thermal desorption as a potentially feasible treatment alternative. The desorption process operates at 650°C under 2.3 atm pressure. The reactor vessel volume is 86 m³ with about 30% of the volume occupied by solids. The volatilized contaminants are present in a 90% nitrogen gas environment at the percentages provided in the table.

component	concentration (% V/V)	molecular weight (g/mol)
contaminant 1	4	78
contaminant 2	2	113
contaminant 3	2	106
contaminant 4	1	92
contaminant 5	1	63
nitrogen gas	90	28

Under normal operating conditions, the extracted gases are contained and treated through air pollution control unit processes. However, under emergency shutdown conditions, the contents of the reactor vessel are directly vented to the atmosphere. The ambient air temperature is 20°C, and the ambient air pressure is 1 atm.

7.1. What is the total mass of the contaminant gas in the reactor vessel?

(A) 180 g

(B) 420 g

(C) 590 g

(D) 840 g

7.2. What is the total volume of the contaminant gas that would be directly emitted to the atmosphere under emergency conditions?

(A) 4.4 m³

(B) 6.0 m³

(C) 8.6 m³

(D) 26 m³

PROBLEM 8

The following question pertain to mobile source air emissions.

8.1. Which state in the United States was the first to adopt legislation for controlling emissions of hydrocarbons and carbon monoxide from motor vehicles?

(A) California

(B) Colorado

(C) Oregon

(D) Massachusetts

8.2. What USEPA test is used for evaluating vehicle emissions and fuel efficiency?

(A) acceleration/deceleration cycle—cold-start (ADCCS) test

(B) cycle vehicle (CVS-75) test

(C) dynamometer load cycle (DY-20) test

(D) power curve/emission rate (PCER) test

8.3. What measures would probably not contribute to overall improved urban air quality?

(A) Reduce the number of miles driven through car-pooling, increasing the use of mass transit, and improving pedestrian and bicycle access.

(B) Provide consumer incentives for the purchase of new cars to increase the ratio of new to used car sales.

(C) Increase the number of diesel-powered vehicles relative to the number of gasoline-powered vehicles through consumer incentives or manufacturer quotas.

(D) Require manufacturers to make zero emission vehicles (ZEV) available to consumers and provide incentives to promote their purchase.

8.4. What is the corporate average fuel economy (CAFE) efficiency of an automobile with city gasoline mileage of 23 mpg and highway mileage of 32 mpg?

(A) 26.33 mpg

(B) 27.05 mpg

(C) 27.21 mpg

(D) 27.50 mpg

PROBLEM 9

The gasoline mileage of an automobile in commuter traffic is 20 mpg. The automobile owner drives 48 mi to and from work (round trip) each day from Monday through Friday. The gasoline chemical formulation is represented by C_8H_{15} with a specific weight of 0.89.

9.1. How many gallons of gasoline are consumed during 1 wk of commuting to and from work?

(A) 12 gal

(B) 17 gal

(C) 24 gal

(D) 34 gal

9.2. What is the required air:fuel ratio for complete combustion of the gasoline to occur?

(A) 3.4

(B) 7.2

(C) 11

(D) 14

9.3. What pollutant will likely increase the most if the fuel:air ratio is less than that required for complete combustion?

(A) carbon monoxide

(B) carbon dioxide

(C) hydrocarbons

(D) nitrous oxides

9.4. If the fuel:air ratio is greater than that required for complete combustion, what mass of carbon dioxide will be emitted by the automobile in 1 wk?

(A) 2 kg/wk

(B) 13 kg/wk

(C) 130 kg/wk

(D) 820 kg/wk

PROBLEM 10

The following questions address issues relating to global climate change.

10.1. Which of the following is not an important greenhouse gas?

(A) H_2O

(B) SO_2

(C) N_2O

(D) O_2

10.2. Why is CO_2 the focus of attention in efforts to control greenhouse gas formation?

(A) CO_2 accounts for over 60% of the radiative forcing caused by historical increases in greenhouse gas concentrations.

(B) CO_2 increases occur with corresponding O_2 decreases that incrementally limit respiration and initiate extinction of sensitive heterotrophic organisms.

(C) In the presence of direct sunlight, CO_2 reacts with free oxygen in the atmosphere to form carbon monoxide and ozone, both of which are greenhouse gases.

(D) Unlike other greenhouse gases, CO_2 can be easily controlled by substituting fuels such as natural gas for fuels such as coal and wood.

10.3. What two factors are believed to most significantly contribute to potential global warming?

(A) combustion of fossil fuels and construction of radiating surfaces such as paved areas and buildings

(B) combustion of fossil fuels and decreased biomass density on the earth's surface

(C) construction of radiating surfaces such as paved areas and buildings and decreased biomass density on the earth's surface

(D) decreased biomass density on the earth's surface and increased atmospheric water vapor from industrialization and agricultural irrigation

Emissions Sources Solutions

SOLUTION 1

1.1. The six contaminants that typically constitute the major components of urban air pollution are NO_x, SO_x, CO, Pb, O_3, and particulate.

The answer is D.

1.2. Incomplete combustion, combustion in air (oxygen is limited to about 20%), and compounds other than C and H in fuels all contribute significantly to pollution from combustion. Moisture existing in fuels and water vapor in air are not significant contributors.

The answer is C.

1.3. Photochemical smog is most typically associated with automobiles. Industrial and sulfurous smog are typically associated with industrial activities.

The answer is A.

1.4. Fossil fuel combustion represents the largest source of air pollution in urban and other industrialized regions of the world.

The answer is C.

SOLUTION 2

2.1. The opacity of black or gray smoke is reported as a Ringlemann number.

The answer is A.

2.2. The opacity of white smoke is reported by percent opacity.

The answer is C.

2.3. Opacity observations are reported to 5% opacity or 1/4 Ringlemann number.

The answer is B.

SOLUTION 3

3.1. The prevailing lapse rate of $-10.1°C/km$ is less than the dry adiabatic lapse rate of $-9.8°C/km$ up to a height of 250 m, indicating superadiabatic conditions. Above 250 m, the prevailing lapse rate has a positive slope and an inversion is formed.

The answer is C.

3.2. Assume the plume cools at the dry adiabatic lapse rate of $-9.8°C/km$ as it rises and the plume will stop rising when the air and plume temperatures are equal. The plume temperature at 250 m is

$$20°C + (250 \text{ m} - 75 \text{ m}) \left(-9.8 \, \frac{°C}{km}\right) \left(\frac{1 \text{ km}}{1000 \text{ m}}\right)$$
$$= 18.3°C$$

The air temperature at 250 m is

$$16°C + (250 \text{ m}) \left(-10.1 \, \frac{°C}{km}\right) \left(\frac{1 \text{ km}}{1000 \text{ m}}\right) = 13.5°C$$

Above 250 m, assume the air temperature will increase at $19.8°C/km$ and the plume temperature will continue to decrease at the dry adiabatic lapse rate.

$$18.3°C + \left(-9.8 \, \frac{°C}{km}\right) \left(\frac{1 \text{ km}}{1000 \text{ m}}\right) x$$
$$= 13.5°C + (19.8°C) \left(\frac{1 \text{ km}}{1000 \text{ m}}\right) x$$

Solve for x.

$$x = \frac{(18.3°C - 13.5°C) \left(1000 \, \frac{m}{km}\right)}{19.8 \, \frac{°C}{km} + 9.8 \, \frac{°C}{km}} = 162 \text{ m}$$

The total height to which the plume will rise is

$$250 \text{ m} + 162 \text{ m} = \boxed{412 \text{ m} \quad (410 \text{ m})}$$

The answer is C.

3.3. When an inversion exists over superadiabatic conditions, a fumigation plume will be formed.

The answer is C.

3.4. The maximum mixing depth occurs at the elevation where the prevailing lapse rate and dry adiabatic lapse rate intersect and is equal to the height to which the plume will rise.

The maximum mixing depth is $\boxed{410 \text{ m.}}$

The answer is C.

3.5. Water vapor in air will condense, and heat will be released as the air rises. Therefore, the dry adiabatic lapse rate will increase (be less negative) in a water-saturated atmosphere.

The answer is A.

SOLUTION 4

4.1. The molecular weight of SO_2 is

$$\text{MW} = 32 \ \frac{\text{g}}{\text{mol}} + (2) \left(16 \ \frac{\text{g}}{\text{mol}}\right) = 64 \text{ g/mol}$$

Assume SO_2 behaves as an ideal gas.

$m = $ mass SO_2, g
$P = $ pressure $= 1$ atm
$V = \dfrac{\text{volume } SO_2}{\text{volume emitted gas}} = \dfrac{0.18 \text{ m}^3}{10^6 \text{ m}^3}$
$R = $ gas law constant $= 8.2 \times 10^{-5}$ m^3·atm/mol·K
$T = $ temperature $= (95°\text{C} + 273)\text{K} = 368\text{K}$
$n = $ number of moles $= \dfrac{\text{mass } SO_2}{\text{MW } SO_2} = \dfrac{m}{\text{MW}}$

$$PV = nRT \text{ or } m = \frac{(\text{MW})PV}{RT}$$

The mass of SO_2 is

$$m = \frac{\left(64 \ \frac{\text{g}}{\text{mol}}\right)(1 \text{ atm})\left(\frac{0.18 \text{ m}^3}{10^6 \text{ m}^3}\right)}{\left(8.2 \times 10^{-5} \ \frac{\text{m}^3 \cdot \text{atm}}{\text{mol} \cdot \text{K}}\right)(368\text{K})} = 382 \text{ g}/10^6 \text{ m}^3$$

The SO_2 concentration is

$$\left(\frac{382 \text{ g}}{10^6 \text{ m}^3}\right)\left(10^6 \ \frac{\mu\text{g}}{\text{g}}\right)$$
$$= \boxed{382 \ \mu\text{g/m}^3 \quad (380 \ \mu\text{g/m}^3)}$$

The answer is C.

4.2. The mass of SO_2 will remain constant as the temperature decreases from 95°C to 18°C, but the volume of the emitted gas will change. The ambient pressure remains constant at 1 atm.

$T_1 = (95°\text{C} + 273)\text{K} = 368\text{K}$
$T_2 = (18°\text{C} + 273)\text{K} = 291\text{K}$
$V_1 = $ volume of gas emitted at T_1
  (given an SO_2 concentration of 0.18 ppm)
  $= 10^6 \text{ m}^3$
$V_2 = $ volume of gas emitted after temperature
  decreases to T_2, m^3
$PV_1 = nRT_1$
$PV_2 = nRT_2$

$$\frac{nR}{P} = \frac{V_1}{T_1} = \frac{V_2}{T_2}$$

$$V_2 = \frac{(10^6 \text{ m}^3)(291\text{K})}{368\text{K}} = 7.9 \times 10^5 \text{ m}^3$$

The SO_2 concentration is

$$\left(\frac{380 \text{ g}}{7.9 \times 10^5 \text{ m}^3}\right)\left(10^6 \ \frac{\mu\text{g}}{\text{g}}\right)$$
$$= \boxed{481 \ \mu\text{g/m}^3 \quad (480 \ \mu\text{g/m}^3)}$$

The answer is A.

4.3. When combined with water, SO_2 will form sulfurous acid according to the following reaction.

$$SO_2 + H_2O \longrightarrow H_2SO_3$$

The answer is C.

SOLUTION 5

5.1. Sulfur trioxide forms sulfuric acid when combined with water. The sulfuric acid subsequently dissociates to hydrogen ion and to sulfate.

$$SO_3 + H_2O \longrightarrow H_2SO_4 \longrightarrow 2H^+ + SO_4^{-2}$$

One mole of SO_3 scrubbed will yield one mole of SO_4^{-2} and two moles of H^+ in the rainwater. The SO_3 concentration in rainwater is

$$\left(\frac{458 \text{ g}}{10^6 \text{ m}^3}\right)\left(\frac{1 \text{ m}^3}{0.27 \text{ m}^3}\right)\left(10^6 \ \frac{\mu\text{g}}{\text{g}}\right)\left(\frac{1 \text{ m}^3}{1000 \text{ L}}\right)$$
$$= 1.7 \ \mu\text{g/L}$$

The molecular weight of SO_3 is

$$32 \ \frac{\text{g}}{\text{mol}} + (3) \left(16 \ \frac{\text{g}}{\text{mol}}\right) = 80 \text{ g/mol}$$

The molecular weight of SO_4^{-2} is

$$32 \ \frac{g}{mol} + (4) \left(16 \ \frac{g}{mol} \right) = 96 \ g/mol$$

The SO_4^{-2} concentration in the rainwater is

$$\frac{\left(1.7 \ \frac{\mu g}{L} \right) \left(\frac{1 \ mol \ SO_4^{-2}}{1 \ mol \ SO_3} \right) \left(96 \ \frac{\mu g}{\mu mol \ SO_4^{-2}} \right)}{80 \ \frac{\mu g}{\mu mol \ SO_3}}$$

$$= \boxed{2.04 \ \mu g/L \quad (2.0 \ \mu g/L)}$$

The answer is B.

5.2. At a pH of 5.5, the H^+ concentration of natural rainwater is

$$10^{-5.5} \ \frac{mol}{L} = 3.16 \times 10^{-6} \ mol/L$$

Two moles of H^+ are added for each mole of SO_4^{-2}. The concentration of H^+ added is

$$\frac{\left(2 \ \frac{mol \ H^+}{mol \ SO_4^{-2}} \right) \left(2.0 \ \frac{\mu g}{L} \right)}{\left(96 \ \frac{\mu g}{\mu mol \ SO_4^{-2}} \right) \left(10^6 \ \frac{\mu mol}{mol} \right)} = 4.17 \times 10^{-8} \ mol/L$$

The final H^+ concentration of the rainwater is

$$3.16 \times 10^{-6} \ \frac{mol}{L} + 4.17 \times 10^{-8} \ \frac{mol}{L} = 3.20 \times 10^{-6} \ \frac{mol}{L}$$

The final pH of the rainwater is

$$-\log \left(3.20 \times 10^{-6} \ \frac{mol}{L} \right) = \boxed{5.49 \quad (5.5)}$$

The answer is C.

SOLUTION 6

6.1. $S + O_2 \longrightarrow SO_2$

The molecular weight of SO_2 is

$$32 \ \frac{g}{mol} + (2) \left(16 \ \frac{g}{mol} \right) = 64 \ g/mol$$

$W = SO_2$ emission rate

$$W = \frac{\left(11 \ \frac{ton \ coal}{h} \right) \left(0.038 \ \frac{ton \ S}{ton \ coal} \right) \left(64 \ \frac{kg \ SO_2}{kmol \ SO_2} \right)}{\left(\frac{1 \ ton}{907 \ kg} \right) \left(32 \ \frac{kg \ S}{kmol \ S} \right) \left(1 \ \frac{kmol \ S}{kmol \ SO_2} \right) \left(3600 \ \frac{s}{h} \right)}$$

$$= \boxed{0.21 \ kg/s}$$

The answer is A.

6.2. σ_z = Gaussian dispersion coefficient for vertical plume concentration at downwind distance x, m

H = effective stack height = 175 m

$$\sigma_z = 0.707H \qquad \begin{bmatrix} \text{applies for maximum} \\ \text{ground level concentration} \end{bmatrix}$$

$$= (0.707)(175 \ m)$$

$$= 124 \ m$$

From the Gaussian dispersion coefficient graph for $\sigma_z = 124$ m and using stability curve C for slightly unstable atmospheric conditions, $x = \boxed{2600 \ m.}$

The answer is A.

6.3. $C_{x,0}$ = concentration, g/m^3, along plume centerline at distance $x = 2600$ m

Q = pollutant emission rate = 210 g/s

μ = average wind speed = 8 m/s

σ_y = Gaussian dispersion coefficient for horizontal plume concentration at downwind distance $x = 2600$ m

$$C_{x,0} = \frac{Q \exp \left((-0.5) \left(\frac{H}{\sigma_z} \right)^2 \right)}{\pi \mu \sigma_z \sigma_y}$$

From the Gaussian dispersion coefficient graph for $x = 2600$ m and using stability curve C for slightly unstable atmospheric conditions, $\sigma_y = 270$ m.

$$C_{2600,0} = \frac{\left(210 \ \frac{g}{s} \right) \exp \left((-0.5) \left(\frac{175 \ m}{124 \ m} \right)^2 \right)}{\pi \left(8 \ \frac{m}{s} \right) (124 \ m)(270 \ m)}$$

$$= 0.000092 \ g/m^3$$

$$\left(0.000092 \ \frac{g}{m^3} \right) \left(10^6 \ \frac{\mu g}{g} \right) = \boxed{92 \ \mu g/m^3}$$

The answer is A.

6.4. At 1500 m and using stability curve C, $\sigma_y = 180$ m and $\sigma_z = 90$ m.

$$C_{1500,0} = \frac{\left(210\,\frac{g}{s}\right)\exp\left((-0.5)\left(\frac{175\text{ m}}{90\text{ m}}\right)^2\right)}{\pi\left(8\,\frac{m}{s}\right)(90\text{ m})(180\text{ m})}$$

$$= 0.000078\text{ g/m}^3$$

$$\left(0.000078\,\frac{g}{m^3}\right)\left(10^6\,\frac{\mu g}{g}\right) = \boxed{78\ \mu g/m^3}$$

The answer is B.

6.5. $C_{x,y}$ = concentration, g/m^3, along plume centerline at distance $x = 1500$ m, and at crosswind distance $y = 200$ m.

$$C_{1500,200} = \frac{\left(\begin{array}{c} Q\exp\left((-0.5)\left(\dfrac{H}{\sigma_z}\right)^2\right) \\[2mm] \times\exp\left((-0.5)\left(\dfrac{y}{\sigma_y}\right)^2\right) \end{array}\right)}{\pi\mu\sigma_z\sigma_y}$$

$$= \frac{\left(\begin{array}{c} \left(210\,\frac{g}{s}\right)\exp\left((-0.5)\left(\dfrac{175\text{ m}}{90\text{ m}}\right)^2\right) \\[2mm] \times\exp\left((-0.5)\left(\dfrac{200\text{ m}}{180\text{ m}}\right)^2\right) \end{array}\right)}{\pi\left(8\,\frac{m}{s}\right)(90\text{ m})(180\text{ m})}$$

$$= 0.000042\text{ g/m}^3$$

$$\left(0.000042\,\frac{g}{m^3}\right)\left(10^6\,\frac{\mu g}{g}\right) = \boxed{42\ \mu g/m^3}$$

The answer is A.

SOLUTION 7

Assume the ideal gas law applies.

P = pressure, atm
V = volume, m^3
n = number of moles
R = gas law constant = 8.2×10^{-5} m^3·atm/mol·K
T = temperature, K

$$PV = nRT$$

Assume Dalton's Law applies.

In a gas mixture, each gas exerts pressure independently of the others and the resulting partial pressure of each gas is proportional to the amount of that gas in the mixture.

7.1. Conditions in the reactor vessel are presented in the following table.

component	concentration (% V/V)	volume (m³)	MW (g/mol)	partial pressure (atm)	moles	mass (g)
contaminant 1	4	2.4	78	0.092	2.92	228
contaminant 2	2	1.2	113	0.046	0.73	82
contaminant 3	2	1.2	106	0.046	0.73	77
contaminant 4	1	0.6	92	0.023	0.18	17
contaminant 5	1	0.6	63	0.023	0.18	11
nitrogen gas	90	54		2.07		
	100	60		2.3		415

$$V = \frac{(86\text{ m}^3)(1-0.3)\left(\frac{\%\text{ V}}{V}\right)}{100\%} = \frac{(60)\left(\frac{\%\text{ V}}{V}\right)\text{ m}^3}{100\%}$$

$$\text{partial pressure} = \frac{(2.3\text{ atm})\left(\frac{\%\text{ V}}{V}\right)}{100\%}$$

$$n = \frac{PV}{RT}$$

$$\text{moles} = \frac{(\text{partial pressure, atm})(\text{volume, m}^3)}{\left(8.2\times10^{-5}\,\frac{\text{m}^3\cdot\text{atm}}{\text{mol·K}}\right)(923\text{K})}$$

$$\text{mass} = (\text{moles})\left(\text{MW},\frac{g}{\text{mol}}\right) = \boxed{415\text{ g}\quad(420\text{ g})}$$

The answer is B.

7.2. Conditions in the ambient air are presented in the following table.

component	concentration (% V/V)	MW (g/mol)	partial pressure (atm)	moles	volume (m³)
contaminant 1	4	78	0.04	2.92	1.75
contaminant 2	2	113	0.02	0.73	0.88
contaminant 3	2	106	0.02	0.73	0.88
contaminant 4	1	92	0.01	0.18	0.43
contaminant 5	1	63	0.01	0.18	0.43
nitrogen gas	90		0.9		
	100		1.0		4.37

$$\text{partial pressure} = \frac{(1.0\text{ atm})\left(\frac{\%\text{ V}}{V}\right)}{100\%}$$

$$V = \frac{nRT}{P} = \frac{(\text{moles})\left(8.2\times10^{-5}\,\frac{\text{m}^3\cdot\text{atm}}{\text{mol·K}}\right)(293\text{K})}{\text{partial pressure, atm}}$$

$$\text{total volume} = \boxed{4.37\text{ m}^3\quad(4.4\text{ m}^3)}$$

The answer is A.

SOLUTION 8

8.1. California was the first state (in 1959) in the United States to adopt legislation for controlling emissions of hydrocarbons and carbon monoxide from motor vehicles.

The answer is A.

8.2. The USEPA test for evaluating vehicle emissions and fuel efficiency is the cycle vehicle (CVS-75) test.

The answer is B.

8.3. Increasing the number of diesel powered vehicles relative to the number of gasoline powered vehicles through consumer incentives or manufacturer quotas would not contribute to overall improved urban air quality.

The answer is C.

8.4. Assume 1 mi of driving with 55% city and 45% highway as prescribed by CAFE.

$$\text{fuel economy} = \frac{1 \text{ mi}}{\dfrac{0.55 \text{ mi}}{23 \, \frac{\text{mi}}{\text{gal}}} + \dfrac{0.45 \text{ mi}}{32 \, \frac{\text{mi}}{\text{gal}}}} = \boxed{26.33 \text{ mpg}}$$

The answer is A.

SOLUTION 9

9.1. $$\frac{\left(48 \, \frac{\text{mi}}{\text{d}}\right)\left(5 \, \frac{\text{d}}{\text{wk}}\right)(1 \text{ wk})}{20 \, \frac{\text{mi}}{\text{gal}}} = \boxed{12 \text{ gal}}$$

The answer is A.

9.2. The molecular weight of C_8H_{15} is

$$(8)\left(12 \, \frac{\text{g}}{\text{mol}}\right) + (15)\left(1 \, \frac{\text{g}}{\text{mol}}\right) = 111 \text{ g/mol}$$

The molecular weight of O_2 is

$$(2)\left(16 \, \frac{\text{g}}{\text{mol}}\right) = 32 \text{ g/mol}$$

The molecular weight of N_2 is

$$(2)\left(14 \, \frac{\text{g}}{\text{mol}}\right) = 28 \text{ g/mol}$$

In air, the N_2:O_2 ratio is 0.75:0.23 by weight.

$$\frac{\left(\dfrac{0.75 \text{ g N}_2}{0.23 \text{ g O}_2}\right)\left(32 \, \dfrac{\text{g O}_2}{\text{mol O}_2}\right)}{\left(28 \, \dfrac{\text{g N}_2}{\text{mol N}_2}\right)} = \frac{3.73 \text{ mol N}_2}{1 \text{ mol O}_2}$$

The stoichiometric air (oxygen and nitrogen) requirement for complete combustion of the gasoline is represented by

$$C_8H_{15} + 11.75O_2 + (3.73)(11.75N_2) \longrightarrow$$
$$8CO_2 + (3.73)(11.75N_2) + 7.5H_2O$$

The mass of C_8H_{15} is

$$(1 \text{ mol})\left(111 \, \frac{\text{g}}{\text{mol}}\right) = 111 \text{ g}$$

The mass of O_2 is

$$(11.75 \text{ mol})\left(32 \, \frac{\text{g}}{\text{mol}}\right) = 376 \text{ g}$$

The mass of N_2 is

$$(3.73)(11.75 \text{ mol})\left(28 \, \frac{\text{g}}{\text{mol}}\right) = 1227 \text{ g}$$

The air:fuel ratio is

$$\frac{376 \text{ g} + 1227 \text{ g}}{111 \text{ g}} = \boxed{14.4 \quad (14)}$$

The answer is D.

9.3. As the air:fuel ratio declines below what is required for complete combustion, the most significant pollutant increase is for carbon monoxide.

The answer is A.

9.4. From Prob. 9.2, 8 moles of CO_2 are produced for each mole of gasoline burned.

The molecular weight of CO_2 is

$$12 \, \frac{\text{g}}{\text{mol}} + (2)\left(16 \, \frac{\text{g}}{\text{mol}}\right) = 44 \text{ g/mol}$$

$$\frac{\left(\begin{array}{c}\left(12 \, \dfrac{\text{gal}}{\text{wk}}\right)\left(3.785 \, \dfrac{\text{L}}{\text{gal}}\right)(0.89) \\ \times \left(1000 \, \dfrac{\text{g}}{\text{L}}\right)\left(44 \, \dfrac{\text{g}}{\text{mol CO}_2}\right)\end{array}\right)}{\left(111 \, \dfrac{\text{g}}{\text{mol C}_8\text{H}_{15}}\right)\left(\dfrac{1 \text{ mol C}_8\text{H}_{15}}{8 \text{ mol CO}_2}\right)\left(1000 \, \dfrac{\text{g}}{\text{kg}}\right)}$$

$$= \boxed{128 \text{ kg/wk} \quad (130 \text{ kg/wk})}$$

The answer is C.

SOLUTION 10

10.1. Important greenhouse gases include H_2O, N_2O, and O_2, but not SO_2.

The answer is B.

10.2. Carbon dioxide (CO_2) is the focus of attention in efforts to control greenhouse gas formation because it accounts for over 60% of the radiative forcing caused by historical increases in greenhouse gas concentrations.

The answer is A.

10.3. The two factors believed to most significantly contribute to potential global warming are combustion of fossil fuels and decreased biomass density on the earth's surface.

The answer is B.

Control Strategies

PROBLEM 1

A countercurrent wet scrubber uses lime to remove HCl from gases generated during primary metal pickling operations. Scrubber design standards and gas characteristics are as follows.

minimum scrubber efficiency for HCl	90%
minimum gas flow velocity	4 m/s
gas flow rate	10 m^3/s
lime feed rate safety factor	5 × stoichiometric
lime slurry concentration	5%
gas HCl concentration	800 ppmv

1.1. What is the minimum diameter required for the scrubber?

   (A) 0.40 m

   (B) 0.71 m

   (C) 1.8 m

   (D) 2.5 m

1.2. What is the daily mass of HCl removed by the scrubber?

   (A) 920 kg/d

   (B) 1100 kg/d

   (C) 2.8×10^5 kg/d

   (D) 7.7×10^5 kg/d

1.3. What is the stoichiometric requirement for lime if available as 100% CaO?

   (A) 720 kg/d

   (B) 890 kg/d

   (C) 4.8×10^5 kg/d

   (D) 19×10^5 kg/d

1.4. What is the actual required slurry feed rate for lime if available as 89% CaO?

   (A) 64 m^3/d

   (B) 80 m^3/d

   (C) 1.2×10^4 m^3/d

   (D) 4.8×10^4 m^3/d

PROBLEM 2

Flares are used to dispose of excess combustible gases including those from landfills and anaerobic biological waste treatment processes.

2.1. What are the hydrogen to carbon ratios, by mass, of methane and acetylene?

   (A) methane H:C = 1:12, acetylene H:C = 1:12

   (B) methane H:C = 1:3, acetylene H:C = 1:12

   (C) methane H:C = 4:1, acetylene H:C = 2:2

   (D) methane H:C = 1:4, acetylene H:C = 1:13

2.2. Does either acetylene or methane produce smoke when flared?

   (A) methane does not because its H:C is close to 1:3, but acetylene does because its H:C is much less

   (B) both will smoke since their H:C is close to 1:12

   (C) neither will smoke since their H:C is greater than 1:12

   (D) methane does because its H:C is greater than 1.0, but acetylene does not because its H:C is less than or equal to 1.0

2.3. What would be the result of injecting steam into the flame zone of an acetylene flare?

   (A) It would extinguish the flame.

   (B) It would trap smoke particulate in condensed steam.

   (C) It would increase mixing with air to improve combustion efficiency.

   (D) It would decrease flame temperature to reduce dioxin formation.

PROBLEM 3

An industrial activity produces a gas flow of 10 m^3/s with the particulate particle size distribution shown in the following table. The average particle specific weight is 2, and the gas temperature is 100°C. A 1.0 m conventional cyclone will be used to control the particulate emissions. Assume a volume shape factor of 0.9.

particle size (μm)	% mass
0–10	10
10–20	34
20–30	43
30–40	13

3.1. What diameter particle will be removed to 100% by the cyclone?

(A) 9 μm

(B) 16 μm

(C) 20 μm

(D) 28 μm

3.2. What is the terminal settling velocity of the particle removed to 100% by the cyclone?

(A) 0.62 m/s

(B) 0.91 m/s

(C) 1.4 m/s

(D) 2.1 m/s

3.3. What is the overall removal efficiency of the cyclone?

(A) 50%

(B) 60%

(C) 70%

(D) 80%

PROBLEM 4

A baghouse is being considered for controlling particulate emissions from a metal parts sandblasting process. The air flow and baghouse parameters include the following.

air flow rate	4 m^3/s
particulate concentration	30 g/m^3
air-cloth ratio	0.01 m^3/m^2·s
filtration velocity	0.01 m/s
fabric resistance	0.003 atm·s/cm
cake resistance	0.02 atm·s·cm/g
bag diameter	20 cm
bag length	2 m

4.1. What is the total area of fabric required?

(A) 260 m^2

(B) 310 m^2

(C) 400 m^2

(D) 480 m^2

4.2. What are the total number of bags required?

(A) 200

(B) 240

(C) 320

(D) 370

4.3. What is the pressure drop through the baghouse following 1 hr of operation?

(A) 0.003 atm

(B) 0.005 atm

(C) 0.02 atm

(D) 0.2 atm

PROBLEM 5

A cosmetics packing process produces a dust-laden air stream at a flow rate of 15 m^3/s. A cross-flow scrubber is being considered for removing the dust from the air stream. The dust has a uniform particle diameter between 0.7 μm and 1.0 μm, and the air stream temperature is 30°C. The following parameters are defined for the scrubber.

gas velocity	40 cm/s
liquid:gas flow ratio	5 × 10^{-4}
fractional target efficiency	0.06
scrubber contact zone length	6 m
liquid droplet diameter	0.03 cm
spray nozzle pressure	0.05 atm

5.1. What is the scrubber efficiency?

(A) 51%

(B) 59%

(C) 64%

(D) 72%

5.2. What is the required cross-sectional area?

(A) 0.40 m^2

(B) 2.7 m^2

(C) 38 m^2

(D) 270 m^2

5.3. What is the air pressure drop through the scrubber spray nozzle?

(A) 0.05 atm

(B) 0.95 atm

(C) 1.0 atm

(D) 1.05 atm

PROBLEM 6

A farm equipment manufacturer has selected electrostatic precipitation to control atmospheric emissions from particulate-producing activities on the paint line. The air flow rate from the line is 8 m^3/s at 42°C. Other parameters are

average electric field	600 000 N/C
average particle diameter	5.0 μm
particulate concentration	12 g/m^3

6.1. What is the particle drift velocity?

- (A) 0.3 m/s
- (B) 0.5 m/s
- (C) 0.7 m/s
- (D) 0.9 m/s

6.2. What plate area is required for 90% efficiency?

- (A) 21 m^2
- (B) 26 m^2
- (C) 37 m^2
- (D) 62 m^2

6.3. What is the daily particulate mass removed at 90% efficiency?

- (A) 310 kg/d
- (B) 630 kg/d
- (C) 950 kg/d
- (D) 7500 kg/d

PROBLEM 7

A dry gas stream contains trichloroethene (TCE) at 0.0016 kg/m^3 and 30 m^3/s flow rate. For the temperature and moisture conditions of the gas stream, the TCE adsorption capacity to granular activated carbon (GAC) is 550 mg/g. Activated carbon adsorbers will have the following characteristics.

gas flow capacity	4 m^3/s
carbon bed diameter	3.5 m
carbon bed capacity	5000 kg

7.1. What is the daily mass of TCE removed?

- (A) 1200 kg/d
- (B) 1600 kg/d
- (C) 2600 kg/d
- (D) 4200 kg/d

7.2. What is the daily mass of activated carbon required to remove the TCE from the gas stream?

- (A) 2200 kg/d
- (B) 3000 kg/d
- (C) 4700 kg/d
- (D) 7700 kg/d

7.3. How many carbon adsorbers are required to treat the gas stream?

- (A) 4
- (B) 6
- (C) 8
- (D) 10

7.4. How frequently will the adsorbers require reactivation?

- (A) 2.6 d
- (B) 5.2 d
- (C) 11 d
- (D) 14 d

PROBLEM 8

A surface condenser is used to control vapor losses from a tank used for acetone storage. The coolant is water with an inlet temperature of 15°C and an outlet temperature of 25°C. The properties of the acetone vapor and the surface condenser include the following.

mass flow rate	5000 kg/h
boiling point	56.2°C
latent heat of vaporization	536 kJ
specific heat	2.15 kJ
maximum outlet temperature	30°C
overall heat transfer coefficient	243 J/$cm^2 \cdot h \cdot °C$

8.1. What is the heat loss from the acetone vapor?

- (A) 2.6×10^5 kJ/d
- (B) 6.3×10^5 kJ/d
- (C) 6.7×10^6 kJ/d
- (D) 7.1×10^7 kJ/d

8.2. What cooling water flow rate is required?

- (A) 4.2 L/min
- (B) 10 L/min
- (C) 130 L/min
- (D) 1200 L/min

8.3. What is the average temperature change in the condenser between the acetone and the water?

    (A) 16°C

    (B) 18°C

    (C) 22°C

    (D) 26°C

8.4. What is the required surface area for the condenser tubes?

    (A) 0.20 m^2

    (B) 0.50 m^2

    (C) 6.0 m^2

    (D) 55 m^2

Control Strategies Solutions

1.1. A = scrubber cross sectional area, m^2
Q = gas flow rate = 10 m^3/s
v = gas flow velocity = 4 m/s

$$A = \frac{Q}{v} = \frac{10\ \frac{m^3}{s}}{4\ \frac{m}{s}} = 2.5\ m^2$$

D = scrubber diameter, m

$$D = \left(\frac{4A}{\pi}\right)^{0.5} = \left(\frac{(4)(2.5\ m^2)}{\pi}\right)^{0.5}$$

$$= \boxed{1.78\ m \quad (1.8\ m)}$$

The answer is C.

1.2. Assume standard temperature and pressure of 25°C and 1 atm, respectively, and assume the ideal gas law applies. 1 mol of HCl gas occupies 24.45 L.

The molecular weight of HCl is

$$1\ \frac{g}{mol} + 35\ \frac{g}{mol} = 36\ g/mol$$

The mass density of the HCl gas is

$$\frac{(800\ m^3)\left(36\ \frac{g}{mol}\right)\left(1000\ \frac{L}{m^3}\right)}{(10^6\ m^3)\left(24.45\ \frac{L}{mol}\right)} = 1.18\ g/m^3$$

The daily mass of removed HCl is

$$\frac{\left(\frac{1.18\ g}{m^3}\right)\left(\frac{10\ m^3}{s}\right)\left(\frac{90\%}{100\%}\right)}{\left(\frac{1\ d}{86\ 400\ s}\right)\left(1000\ \frac{g}{kg}\right)}$$

$$= \boxed{918\ kg/d \quad (920\ kg/d)}$$

The answer is A.

1.3. $2HCl + CaO \longrightarrow Ca^{+2} + 2Cl^- + H_2O$

For 2 mol of HCl, 1 mol of CaO is required.

The molecular weight of CaO is

$$40\ \frac{g}{mol} + 16\ \frac{g}{mol} = 56\ g/mol$$

The daily mass requirement of CaO is

$$\frac{\left(920\ \frac{kg}{d}\right)\left(56\ \frac{kg}{kmol\ CaO}\right)}{\left(36\ \frac{kg}{kmol\ HCl}\right)\left(\frac{2\ kmol\ HCl}{1\ kmol\ CaO}\right)}$$

$$= \boxed{715\ kg/d \quad (720\ kg/d)}$$

The answer is A.

1.4. Assume slurry density is the same as water density, 1000 kg/m^3. The slurry feed rate is

$$\frac{(5)\left(715\ \frac{kg\ CaO}{d}\right)\left(\frac{1\ m^3}{1000\ kg}\right)}{\left(\frac{89\%}{100\%}\right)\left(\frac{5\ kg\ CaO}{100\ kg\ slurry}\right)} = \boxed{80\ m^3/d}$$

The answer is B.

2.1. The chemical formula of methane is CH$_4$.

$$(1\ mol\ C)\left(12\ \frac{g}{mol}\right) = 12\ g\ C$$

$$(4\ mol\ H)\left(1\ \frac{g}{mol}\right) = 4\ g\ H$$

$$H:C = 4:12 = 1:3$$

The chemical formula of acetylene is C_2H_2.

$$(2 \text{ mol C}) \left(12 \frac{g}{mol}\right) = 24 \text{ g C}$$

$$(2 \text{ mol H}) \left(1 \frac{g}{mol}\right) = 2 \text{ g H}$$

$$H{:}C = 2{:}24 = \boxed{1{:}12}$$

The answer is B.

2.2. A flare will typically burn smokeless when H:C is 1:3 or greater. Therefore, methane with H:C = 1:3 will likely burn smokeless and acetylene with H:C = 1:12 will likely burn smoky.

The answer is A.

2.3. Injecting steam into the flame zone of the acetylene flare will increase turbulence to promote better mixing with oxygen. This reduces the smokiness and flame length of an acetylene flare.

The answer is C.

SOLUTION 3

3.1. g = centrifugal acceleration, m/s
Q = gas flow rate = 10 m³/s
D = cylinder diameter = 1.0 m
D_e = outlet diameter = $0.5D$ = 0.5 m
L_1 = cylinder length = $2D$ = 2.0 m
L_2 = cone length = $2D$ = 2.0 m
r_a = average radius at particle location

$$\begin{aligned} r_a &= r_e + 0.5(r - r_e) \\ &= 0.5D_e + (0.5)(0.5D - 0.5D_e) \\ &= 0.375 \text{ m} \end{aligned}$$

$$g = \frac{\left(\dfrac{4Q}{(D - D_e)(2L_1 + L_2)}\right)^2}{r_a}$$

$$= \frac{\left(\dfrac{(4)(10 \frac{m^3}{s})}{(1.0 \text{ m} - 0.5 \text{ m})((2)(2.0 \text{ m}) + 2.0 \text{ m})}\right)^2}{0.375 \text{ m}}$$

$$= 474 \text{ m/s}^2$$

d_p = diameter particle removed to 100%, m
μ = gas viscosity = 2.2×10^{-5} kg/m·s for air at 100°C
ρ_p = particle density = 2000 kg/m³

$$d_p = (4) \left(\frac{Q\mu}{g\rho_p(2L_1 + L_2)(D + D_e)}\right)^{0.5}$$

$$= (4) \left(\frac{\left(10 \frac{m^3}{s}\right)\left(2.2 \times 10^{-5} \frac{kg}{m \cdot s}\right)}{\begin{array}{c}\left(474 \frac{m}{s^2}\right)\left(2000 \frac{kg}{m^3}\right) \\ \times \left((2)(2.0 \text{ m}) + 2.0 \text{ m}\right) \\ \times (1.0 \text{ m} + 0.5 \text{ m})\end{array}}\right)^{0.5}$$

$$= 2.0 \times 10^{-5} \text{ m}$$

$$= (2.0 \times 10^{-5} \text{ m}) \left(10^6 \frac{\mu m}{m}\right)$$

$$= \boxed{20 \ \mu m}$$

The answer is C.

3.2. v_t = terminal settling velocity of the 100% removed particle, m/s

$$v_t = \frac{Q}{0.5\pi(L_1 + 0.5L_2)(D + D_e)}$$

$$= \frac{10 \frac{m^3}{s}}{0.5\pi(2.0 \text{ m} + (0.5)(2.0 \text{ m}))(1.0 \text{ m} + 0.5 \text{ m})}$$

$$= \boxed{1.4 \text{ m/s}}$$

The answer is C.

3.3. v_i = particle terminal settling velocity for each increment, m/s
$d_i = 1.24\beta^{1/3}d_{pi}$
= average diameter of the equivalent spherical particle for each increment, m
β = volume shape factor = 0.9
ρ_a = gas density = 1.15 kg/m³ for air at 100°C

$$v_i = \left(\frac{4g\rho_p d_i}{3C_D \rho_a}\right)^{0.5}$$

$$= \left(\frac{(4)\left(474 \frac{m}{s^2}\right)\left(2000 \frac{kg}{m^3}\right)(1.24)\left(0.9^{1/3}\right)d_{pi}}{3C_D\left(1.15 \frac{kg}{m^3}\right)}\right)^{0.5}$$

$$= (1147)\left(\frac{d_{pi}}{C_D}\right)^{0.5}$$

Re = Reynolds number, dimensionless

$$Re = \frac{d_i v_i \rho_a}{\mu} = \frac{v_i(1.24)(0.9^{1/3})d_{pi}\left(1.15\ \frac{kg}{m^3}\right)}{2.2 \times 10^{-5}\ \frac{kg}{m \cdot s}}$$

$$= 62\,581 v_i d_{pi}\ \frac{s}{m^2}$$

C_D = drag coefficient

$$C_D = \frac{24}{Re} = \frac{24}{62\,581 v_i d_{pi}\ \frac{s}{m^2}} = \frac{3.8 \times 10^{-4}\ \frac{m^2}{s}}{v_i d_{pi}}$$

Substituting the expression for C_D into the equation for v_i yields the following.

$$v_i = (1147)\left(\frac{d_{pi}}{C_D}\right)^{0.5} = (1147)\left(\frac{v_i d_{pi}^2}{3.8 \times 10^{-4}\ \frac{m^2}{s}}\right)^{0.5}$$

$$= (3.4 \times 10^9)d_{pi}^2$$

particle size (μm)	mass %	cumulative mass fraction	$d_{pi}(10^{-6}\ m)$	v_i (m/s)
0–10	10	0.10	5	0.085
10–20	34	0.44	15	0.77
20–30	43	0.87	25	2.1
30–40	13	1.0	35	4.2
	100			

The following figure presents a plot of cumulative mass fraction and incremental settling velocity. Cumulative mass fraction is represented by x.

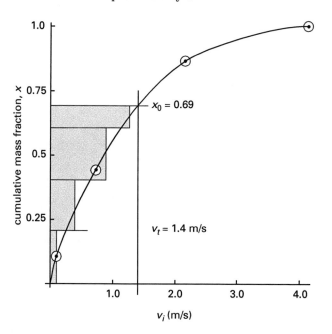

The following table summarizes the incremental mass fraction, Δx, and the corresponding incremental settling velocity resulting from the integration of the plot presented in the figure.

Δx	v_i (m/s)	$\Delta x v_i$ (m/s)
0.2	0.085	0.017
0.2	0.4	0.08
0.2	0.9	0.18
0.095	1.28	0.12
		0.40

E = overall efficiency
x_o = fraction removed corresponding to v_t
    = 0.69 (from the figure)

$$E = (1 - x_o) + \frac{\Sigma \Delta x v_i}{v_t} = (1 - 0.69) + \frac{0.40\ \frac{m}{s}}{1.4\ \frac{m}{s}}$$

$$= 0.596 = \boxed{59.6\%\quad(60\%)}$$

The answer is B.

SOLUTION 4

4.1. Q_g = air flow rate = 4 m³/s
        A/C = air:cloth ratio = 0.01 m³/m²·s

The total fabric area is

$$\frac{Q_g}{\frac{A}{C}} = \frac{4\ \frac{m^3}{s}}{0.01\ \frac{m^3}{m^2 \cdot s}} = \boxed{400\ m^2}$$

The answer is C.

4.2. The area/bag is

$$\pi(\text{bag diameter})(\text{bag length})$$

$$= \pi(20\ cm)(2\ m)\left(\frac{1\ m}{100\ cm}\right)$$

$$= 1.26\ m^2$$

The number of bags is

$$\frac{400\ m^2}{1.26\ m^2} = \boxed{317.5\quad(320)}$$

The answer is C.

4.3. $-\Delta P$ = pressure drop, atm
        K_1 = fabric resistance = 0.003 atm/(cm/s)
        v_f = filtration velocity = 0.01 m/s
        K_2 = cake resistance
            = 0.02 atm/(cm/s)(g/cm²)
        C_p = particulate concentration = 30 g/m³
        t = operating period = 1 h

$$-\Delta P = K_1 v_f + K_2 C_p v_f^2 t$$

$$= \left(0.003 \frac{\text{atm}}{\left(\frac{\text{cm}}{\text{s}}\right)}\right) \left(0.01 \frac{\text{m}}{\text{s}}\right) \left(100 \frac{\text{cm}}{\text{m}}\right)$$

$$+ \frac{\left(\begin{array}{c}(0.02 \text{ atm})\left(30 \frac{\text{g}}{\text{m}^3}\right) \\ \times \left(0.01 \frac{\text{m}}{\text{s}}\right)^2 (1 \text{ h})\end{array}\right)}{\left(1 \frac{\text{cm}}{\text{s}}\right)\left(1 \frac{\text{g}}{\text{cm}^2}\right)\left(100 \frac{\text{cm}}{\text{m}}\right)\left(\frac{1 \text{ h}}{3600 \text{ s}}\right)}$$

$$= \boxed{0.005 \text{ atm}}$$

The answer is B.

SOLUTION 5

5.1. Z_o = scrubber contact zone length = 6 m
E_f = fractional target efficiency = 0.06
d_d = liquid droplet diameter = 0.03 cm
Q_l/Q_g = liquid:gas flow ratio = 5×10^{-4}

$$k\Delta P = \left(\frac{3Z_o E_f}{2d_d}\right)\left(\frac{Q_l}{Q_g}\right)$$

$$= \left(\frac{(3)(6 \text{ m})(0.06)}{(2)(0.03 \text{ cm})\left(\frac{1 \text{ m}}{100 \text{ cm}}\right)}\right)(5 \times 10^{-4})$$

$$= 0.9$$

The overall percent efficiency is

$$E\% = (100\%)(1 - e^{-k\Delta P})$$
$$= (100\%)(1 - e^{-0.9})$$
$$= \boxed{59.3\% \quad (59\%)}$$

The answer is B.

5.2. A_x = cross-sectional area, m^2
Q_g = gas flow rate = 15 m^3/s
v_g = gas velocity = 40 cm/s

$$A_x = \frac{Q_g}{v_g} = \frac{15 \frac{\text{m}^3}{\text{s}}}{\left(40 \frac{\text{cm}}{\text{s}}\right)\left(\frac{1 \text{ m}}{100 \text{ cm}}\right)}$$

$$= \boxed{37.5 \text{ m}^2 \quad (38 \text{ m}^2)}$$

The answer is C.

5.3. The pressure drops from 0.05 atm at the nozzle inlet to zero as a free jet at the nozzle outlet. The pressure drop is

$$0.05 \text{ atm} - 0.0 \text{ atm} = \boxed{0.05 \text{ atm}}$$

The answer is A.

SOLUTION 6

6.1. w = drift velocity, m/s
ϵ_o = permittivity constant
$= 8.85 \times 10^{-12}$ C^2/N·m^2
E = average electric field = 600 000 N/C
d_p = average particle diameter = 5.0 μm
μ_g = gas viscosity
$= 1.8 \times 10^{-5}$ kg/m·s for air at 42°C

$$w = \frac{8\epsilon_o E^2 d_p}{24\mu_g}$$

$$= \frac{\left(\begin{array}{c}(8)\left(8.85 \times 10^{-12} \frac{\text{C}^2}{\text{N·m}^2}\right) \\ \times \left(600\,000 \frac{\text{N}}{\text{C}}\right)^2 (5.0 \ \mu\text{m})\end{array}\right)}{(24)\left(1.8 \times 10^{-5} \frac{\text{kg}}{\text{m·s}}\right)\left(1 \frac{\text{N·s}^2}{\text{kg·m}}\right)\left(10^6 \frac{\mu\text{m}}{\text{m}}\right)}$$

$$= \boxed{0.295 \text{ m/s} \quad (0.3 \text{ m/s})}$$

The answer is A.

6.2. A = plate area, m^2
$E\%$ = efficiency = 90%
Q_g = gas flow rate = 8 m^3/s

$$A = \frac{-\ln\left(1 - \left(\frac{E\%}{100\%}\right)\right)Q_g}{w}$$

$$= \frac{-\ln\left(1 - \left(\frac{90\%}{100\%}\right)\right)\left(8 \frac{\text{m}^3}{\text{s}}\right)}{0.3 \frac{\text{m}}{\text{s}}}$$

$$= \boxed{61.4 \text{ m}^2 \quad (62 \text{ m}^2)}$$

The answer is D.

6.3. C_p = particulate concentration = 12 g/m³

The mass removed is

$$\left(\frac{E\%}{100\%}\right)C_pQ_g = \frac{\left(\frac{90\%}{100\%}\right)\left(12\ \frac{g}{m^3}\right)\left(8\ \frac{m^3}{s}\right)}{\left(\frac{1\ d}{86\,400\ s}\right)\left(1000\ \frac{g}{kg}\right)}$$

$$= \boxed{7465\ \text{kg/d} \quad (7500\ \text{kg/d})}$$

The answer is D.

SOLUTION 7

7.1. The TCE mass flow rate is

$$\left(30\ \frac{m^3}{s}\right)\left(0.0016\ \frac{kg}{m^3}\right) = 0.048\ \text{kg/s}$$

Assume 100% adsorption efficiency.

The TCE removed is

$$\left(0.048\ \frac{kg}{s}\right)\left(86\,400\ \frac{s}{d}\right) = \boxed{4147\ \text{kg/d} \quad (4200\ \text{kg/d})}$$

The answer is D.

7.2. The mass of GAC required is

$$\frac{\left(4200\ \frac{kg\ TCE}{d}\right)\left(10^6\ \frac{mg}{kg}\right)}{\left(550\ \frac{mg\ TCE}{g\ GAC}\right)\left(1000\ \frac{g}{kg}\right)}$$

$$= \boxed{7636\ \text{kg/d} \quad (7700\ \text{kg/d})}$$

The answer is D.

7.3. The number of adsorbers is

$$\frac{30\ \frac{m^3}{s}}{4\ \frac{m^3}{s}} = \boxed{7.5 \quad (8)}$$

The answer is C.

7.4. The total GAC mass in 8 adsorbers is

$$(8)(5000\ \text{kg}) = 40\,000\ \text{kg}$$

$$\frac{40\,000\ \text{kg}}{7700\ \frac{kg}{d}} = \boxed{5.2\ \text{d}}$$

The answer is B.

SOLUTION 8

8.1. q_a = heat, cal
$\dot{m}$ = acetone mass rate = 5000 kg/h
C_a = acetone specific heat = 2.15 kJ/kg·°C
T_{1a} = acetone initial temperature = 56.2°C
T_{2a} = acetone final temperature = 30°C
H_v = latent heat of vaporization = 536 kJ/kg

$$q_a = \dot{m}C_a(T_{1a} - T_{2a}) + \dot{m}H_v$$

$$= \left(5000\ \frac{kg}{h}\right)\left(2.15\ \frac{kJ}{kg\cdot°C}\right)(56.2°C - 30°C)\left(24\ \frac{h}{d}\right)$$

$$+ \left(5000\ \frac{kg}{h}\right)\left(536\ \frac{kJ}{kg}\right)\left(24\ \frac{h}{d}\right)$$

$$= \boxed{7.1 \times 10^7\ \text{kJ/d}}$$

The answer is D.

8.2. Q_w = cooling water flow rate, L/min
T_{1w} = water final temperature = 25°C
T_{2w} = water initial temperature = 15°C
C_w = water specific heat = 4.184 kJ/kg·°C
ρ_w = water density = 1 kg/L

$$Q_w = \frac{q_a}{C_w(T_{1w} - T_{2w})\rho_w}$$

$$= \frac{\left(7.1 \times 10^7\ \frac{kJ}{d}\right)\left(\frac{1\ d}{1440\ min}\right)}{\left(4.184\ \frac{kJ}{kg\cdot°C}\right)(25°C - 15°C)\left(\frac{1\ kg}{L}\right)}$$

$$= \boxed{1181\ \text{L/min} \quad (1200\ \text{L/min})}$$

The answer is D.

8.3.
$$\Delta T = \frac{(T_{1a} - T_{1w}) - (T_{2a} - T_{2w})}{\ln\left(\dfrac{T_{1a} - T_{1w}}{T_{2a} - T_{2w}}\right)}$$

$$= \frac{(56.2°C - 25°C) - (30°C - 15°C)}{\ln\left(\dfrac{56.2°C - 25°C}{30°C - 15°C}\right)}$$

$$= \boxed{22.1°C \quad (22°C)}$$

The answer is C.

8.4. A_s = tube surface area, m^2
U = overall heat transfer coefficient
$\quad$ = 243 J/cm²·h·°C
$$A_s = \frac{q_a}{U \Delta T}$$

$$= \frac{\left(7.1 \times 10^7\ \dfrac{kJ}{d}\right)\left(1000\ \dfrac{J}{kJ}\right)\left(\dfrac{1\ d}{24\ h}\right)}{\left(243\ \dfrac{J}{cm^2 \cdot h \cdot °C}\right)(22°C)\left(10^4\ \dfrac{cm^2}{m^2}\right)}$$

$$= \boxed{55.3\ m^2 \quad (55\ m^2)}$$

The answer is D.

Section III
Solid, Hazardous, and Special Waste

- Solid Waste and Hazardous Waste

Solid Waste and Hazardous Waste

PROBLEM 1

A city is considering alternatives for the management of solid waste generated by its residents. The solid waste characteristics for the city are as follows.

component	% mass	component discarded % moisture	component discarded dry density (kg/m^3)	component discarded dry energy (kJ/kg)
paper	44	6	85	16 750
garden	17	60	105	6500
food	11	70	290	4650
cardboard	9	5	50	16 300
wood	7	20	240	18 600
plastic	7	2	65	32 600
miscellaneous inert materials	5	8	480	7000

The per capita solid waste generation rate for the 82,000 residents of the city is 2.7 kg/d.

1.1. What is the moisture content of the bulk waste as discarded?

(A) 23 kg/100 kg

(B) 68 kg/100 kg

(C) 76 kg/100 kg

(D) 88 kg/100 kg

1.2. What is the bulk density of the waste when dry?

(A) 26 kg/m^3

(B) 68 kg/m^3

(C) 88 kg/m^3

(D) 115 kg/m^3

1.3. What is the daily energy content available from the waste when dry?

(A) 2.2×10^3 kJ/d

(B) 2.2×10^5 kJ/d

(C) 2.8×10^9 kJ/d

(D) 2.8×10^{11} kJ/d

1.4. Is the energy content of the waste adequate to dry the waste without having to add energy from an external source?

(A) Yes, the energy content of the waste is greater than the energy required to dry it.

(B) No, the energy content of the waste is less than the energy required to dry it.

(C) Yes, the energy content of the waste is less than the energy required to dry it.

(D) No, the energy content of the waste is greater than the energy required to dry it.

PROBLEM 2

The chemical characterization of a municipal solid waste is summarized in the following table. The chemical characterization is based on typical published values for the waste components listed.

waste component	dry mass (kg/100 kg)	dry elemental chemical composition (%)					
		C	H	O	N	S	ash
food	4.9	48.0	6.4	37.6	2.6	0.4	5.0
glass/metal	3.2	–	–	–	–	–	100
paper	12.6	43.5	6.0	44.0	0.3	0.2	6.0
plastic	8.7	60.0	7.2	22.8	–	–	10
wood debris	2.1	49.5	6.0	42.7	0.2	0.1	1.5
yard clippings	29.5	47.8	6.0	38.0	3.4	0.3	4.5

2.1. What is the chemical formula of the waste if sulfur is included?

(A) $C_{50}H_{132}O_{61}N_5S$

(B) $C_{84}H_{265}O_{93}N_{12}S$

(C) $C_{259}H_{848}O_{376}N_{16}S$

(D) $C_{523}H_{1795}O_{795}N_{20}S$

2.2. What is the chemical formula of the waste if sulfur is excluded?

(A) $C_5H_{32}O_6N$

(B) $C_{11}H_{45}O_{18}N$

(C) $C_{27}H_{92}O_{41}N$

(D) $C_{50}H_{132}O_{61}N$

2.3. What is the ash-free energy content of the waste?

(A) 2100 kJ/kg

(B) 12 000 kJ/kg

(C) 110 000 kJ/kg

(D) 240 000 kJ/kg

PROBLEM 3

The residents of a city with a population of 25,000 generate solid waste at a rate of 1.85 kg/d per capita. The discarded moisture content of the waste is 32%, and the chemical formula of the waste is $C_{40}H_{86}O_{37}N$. The waste is landfilled, and the city collects the landfill gases and flares them to the atmosphere. The city manager sees an opportunity to offset expenses by using methane gas from the landfill to replace natural gas purchases. The city buys natural gas at $0.21/\text{m}^3$.

3.1. What is the annual total mass of solid waste generated by the city's population?

(A) 1.2×10^2 kg/yr

(B) 4.6×10^4 kg/yr

(C) 4.9×10^6 kg/yr

(D) 1.7×10^7 kg/yr

3.2. What is the annual total volume of methane gas, at 1 atmosphere pressure and 25°C, potentially produced from decomposition of the waste?

(A) 5.4×10^4 m^3/yr

(B) 7.3×10^6 m^3/yr

(C) 2.1×10^7 m^3/yr

(D) 7.7×10^9 m^3/yr

3.3. What are the potential annual savings, ignoring all other costs, if natural gas purchases are offset by methane gas recovered from the landfill?

(A) $710,000/yr

(B) $1,500,000/yr

(C) $4,400,000/yr

(D) $1,600,000,000/yr

3.4. Is the methane gas useable as a replacement fuel for natural gas directly upon recovery from the landfill?

(A) yes

(B) No, the gas needs to be blended with propane or butane to increase its energy value.

(C) No, the gas needs to be dried to remove water vapor and scrubbed to remove gases that are noncombustible.

(D) No, the gas needs to be odorized and stabilized by blending with nitrogen gas.

PROBLEM 4

A landfill is sited in a natural low-permeability clay 4.0 m thick with a hydraulic conductivity of 10^{-8} cm/s. Chloride has been selected as a tracer to evaluate diffusion through the clay. The chloride concentration is 1106 mg/L with a diffusion coefficient of 2.03×10^{-9} m^2/s and a tortuosity coefficient of 0.06.

4.1. What is the value of the apparent (effective) diffusion coefficient for chloride in the clay?

(A) 3.4×10^{-8} m^2/s

(B) 2.0×10^{-9} m^2/s

(C) 1.2×10^{-10} m^2/s

(D) 2.6×10^{-11} m^2/s

4.2. How long will it take for the chloride to diffuse through the clay to a 10 mg/L concentration?

(A) 3.2 yr

(B) 32 yr

(C) 320 yr

(D) 3200 yr

PROBLEM 5

28% of municipal solid waste collected from a population of 125,000 where the per-capita generation rate is 1.6 kg/d is recovered or recycled. The remaining waste is landfilled. The landfilled waste is compacted to a maximum in-place density of 850 kg/m^3 and the soil cover-to-compacted waste ratio is 1:5 by volume. The operating life of the landfill is 30 yr.

5.1. What is the annual mass of waste requiring disposal in the landfill?

(A) 2.0×10^7 kg/yr

(B) 5.3×10^7 kg/yr

(C) 1.0×10^8 kg/yr

(D) 2.6×10^8 kg/yr

5.2. What is the annual in-place volume of the waste requiring disposal in the landfill?

(A) 2.4 ha·m/yr

(B) 6.2 ha·m/yr

(C) 12 ha·m/yr

(D) 31 ha·m/yr

5.3. What is the annual volume of soil cover required for the landfill?

(A) 0.48 ha·m/yr

(B) 0.70 ha·m/yr

(C) 1.2 ha·m/yr

(D) 6.2 ha·m/yr

5.4. What is the total volume required for the landfill over its operating life?

(A) 86 ha·m

(B) 220 ha·m

(C) 430 ha·m

(D) 1100 ha·m

PROBLEM 6

Four alternative sites are being evaluated for construction of a RCRA-permitted hazardous waste landfill. The site characteristics are summarized in the following table.

category	criteria	WF[1]	site A R[2]	site B R	site C R	site D R
soil	permeability	1	4	3	5	2
	heterogeneities	3	2	2	1	3
geology	seismic activity	3	5	5	5	4
groundwater	quality/use	7	6	4	3	5
	gradient/depth	5	7	4	4	3
hydrology	topography	7	3	5	2	4
	streams/lakes	4	4	3	6	4
community	population	3	6	7	3	3
	land uses	7	7	7	2	5

[1] weighting factor, scaled from 1 (more important) to 7 (less important)

[2] rating, scaled from 1 (more satisfactory) to 7 (less satisfactory)

6.1. Which site best meets the desired criteria?

(A) site A

(B) site B

(C) site C

(D) site D

6.2. Within the borders of which areas are landfills not prohibited?

(A) wetlands

(B) critical habitat areas

(C) U.S. military installations

(D) 100 yr flood plains

6.3. What common source provides soils information to assist in landfill siting assessments?

(A) U.S. Bureau of Indian Affairs

(B) local departments of public health

(C) U.S. Army Corps of Engineers

(D) USDA Soil Conservation Service

6.4. What criteria is typically most critical in landfill siting?

(A) economics

(B) public acceptance

(C) technical feasibility

(D) environmental protection

PROBLEM 7

The following figure presents a schematic of a landfill and cap, and the accompanying table summarizes the cap and drainage layer material, fill thicknesses, and hydraulic conductivities. The maximum desired leachate head is 30 cm and the leachate collection laterals are 150 m long and are placed on the bottom of the drainage layer at 25 m intervals. Assume a self-cleaning velocity on the laterals of 0.6 m/s.

material	thickness (cm)	hydraulic conductivity (cm/s)
top soil	30	10^{-3}
drainage sub-layer	15	10^{-2}
clay cap layer	60	10^{-6}
waste fill	3200	10^{-3}
drainage layer	30	10^{-1}

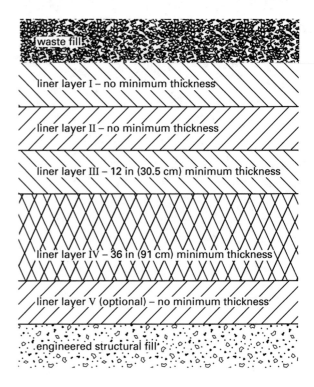

7.1. What is the overall hydraulic conductivity for all layers, including the waste fill?

(A) 1.7×10^{-6} cm/s

(B) 5.3×10^{-5} cm/s

(C) 580 cm/s

(D) 980 cm/s

7.2. What is the required spacing of the leachate collection laterals in the drainage layer?

(A) 6.0 m

(B) 26 m

(C) 36 m

(D) 43 m

7.3. What is the leachate flow rate to each lateral in the drainage layer?

(A) 6.8×10^{-6} m^3/s

(B) 1.8×10^{-4} m^3/s

(C) 1.0×10^{-3} m^3/s

(D) 2.7×10^{-2} m^3/s

7.4. What is the required diameter of the laterals?

(A) 2.3 cm

(B) 4.7 cm

(C) 40 cm

(D) 82 cm

PROBLEM 8

The following figure presents a schematic of a RCRA Subtitle C landfill liner system.

8.1. Which liner layer represents the leachate collection, detection, and removal system (LCDRS)?

(A) liner layer I

(B) liner layer II

(C) liner layer III

(D) liner layer IV

8.2. Which liner layer represents the geomembrane?

(A) liner layer I

(B) liner layer II

(C) liner layer III

(D) liner layer IV

8.3. Which liner layer represents the soil liner?

(A) liner layer I

(B) liner layer II

(C) liner layer III

(D) liner layer IV

8.4. Which liner layer represents the leachate collection and removal system (LCRS)?

(A) liner layer I

(B) liner layer II

(C) liner layer III

(D) liner layer IV

PROBLEM 9

A landfill located in an arid region of the United States has an in-place waste capacity of 3.0×10^8 kg. The landfill has been closed for 10 yr. The methane gas production decay rate constant is estimated at 0.05 yr^{-1}.

9.1. What type of gas venting would not be appropriate for a landfill that accepted hazardous waste?

(A) surface-exposed gravel cell or trench vent

(B) surface-sealed gravel cell or trench vent

(C) vertical vapor well

(D) vented perforated pipe

9.2. What is the likely composition of the gas produced by the landfill?

(A) 15% CO_2, 25% CH_4, and 60% others

(B) 20% CO_2, 35% CH_4, and 45% others

(C) 30% CO_2, 40% CH_4, and 30% others

(D) 40% CO_2, 60% CH_4, and <1% others

9.3. What is the approximate current average rate at which gas is produced by the landfill?

(A) 6.5×10^2 m^3/yr

(B) 2.5×10^3 m^3/yr

(C) 9.3×10^5 m^3/yr

(D) 1.8×10^9 m^3/yr

9.4. What will be the approximate average rate at which gas is produced by the landfill 15 yr in the future?

(A) 3.1×10^2 m^3/yr

(B) 1.2×10^3 m^3/yr

(C) 4.4×10^5 m^3/yr

(D) 8.5×10^8 m^3/yr

PROBLEM 10

Municipal solid waste is being evaluated for composting with thickened waste activated sludge. The city generates 300 000 kg of solid waste daily, 32% of which is compostable. The discarded moisture content of the solid waste is 38%, and the waste C:N ratio is 12:1. The waste activated sludge is thickened to 18% solids, and the C:N ratio of the solids is 90:1.

10.1. How much sludge must be mixed with the solid waste to produce a C:N ratio of 30:1?

(A) 22 000 kg/d

(B) 29 000 kg/d

(C) 220 000 kg/d

(D) 290 000 kg/d

10.2. If the C:N ratio of 30:1 is satisfied by the sludge, what is the resulting moisture content of the mixture?

(A) 24%

(B) 30%

(C) 56%

(D) 69%

10.3. How much water must be added or removed to bring the moisture content of the sludge-solid waste mixture, blended to meet the C:N ratio of 30:1, to 60%?

(A) add 120 000 kg/d

(B) add 99 000 kg/d

(C) add 13 000 kg/d

(D) remove 28 000 kg/d

10.4. What is the optimum operating temperature range for composting?

(A) 25°C to 30°C

(B) 35°C to 40°C

(C) 45°C to 50°C

(D) 55°C to 60°C

10.5. What classification best describes composting?

(A) mesophilic aerobic

(B) mesophilic anaerobic

(C) thermophylic aerobic

(D) thermophylic anaerobic

PROBLEM 11

Injection wells are used for a variety of purposes in the United States. The following problems address these uses.

11.1. What class of injection well is used for injection of hazardous wastes?

(A) Class I

(B) Class II

(C) Class III

(D) Class IV

11.2. What class of injection well is used for aquifer recharge?

(A) Class I

(B) Class II

(C) Class IV

(D) Class V

11.3. What class of injection well is typically associated with oil and gas production dry holes?

(A) Class I

(B) Class II

(C) Class III

(D) Class V

11.4. What injection well use does not require injection below any formation that has an underground source of drinking water within $1/4$ mi of the bore hole?

(A) any use involving a Class I injection well

(B) any use involving a Class II injection well

(C) any use involving a Class IV injection well

(D) only uses where groundwater recharge is the objective of injection

PROBLEM 12

A hazardous waste incinerator receives waste on a continuous basis at 1800 kg/h. The principal organic hazardous constituent (POHC) makes up 27% by mass of the waste mixture. Stack monitoring results show CO_2 emissions at 1600 kg/h, CO emissions at 0.13 kg/h, and POHC emissions at 0.0433 kg/h. Particulate is measured at 203 mg/m^3.

12.1. What is the incinerator combustion efficiency?

(A) 88.8888%

(B) 99.9911%

(C) 99.9919%

(D) 99.9973%

12.2. What is the incinerator destruction and removal efficiency (DRE) for the POHC?

(A) 88.8888%

(B) 99.9911%

(C) 99.9919 %

(D) 99.9973%

12.3. What is the particulate concentration corrected to 4% O_2?

(A) 110 mg/m^3

(B) 140 mg/m^3

(C) 170 mg/m^3

(D) 180 mg/m^3

PROBLEM 13

An organic solvent waste is to be incinerated at a waste feed rate of 1200 L/h, 24 h/d. The waste contains 20% water and 80% organic solvent. The specific gravity of the waste is 0.95. The composition of the organic solvent fraction is provided in the following table.

organic solvent component	weight %
carbon	78
hydrogen	10
oxygen	10
chloride	2

13.1. What is the waste feed rate?

(A) 60 kg/h

(B) 1140 kg/h

(C) 1260 kg/h

(D) 24 000 kg/h

13.2. Assuming complete combustion, what are the mass flows of CO_2 and HCl produced as combustion products?

(A) CO_2 = 710 kg/h, HCl = 23 kg/h

(B) CO_2 = 920 kg/h, HCl = 980 kg/h

(C) CO_2 = 2600 kg/h, HCl = 19 kg/h

(D) CO_2 = 3300 kg/h, HCl = 25 kg/h

13.3. What is the stoichiometric mass flow requirement of O_2 for complete combustion of the carbon in the waste?

(A) 710 kg/h

(B) 940 kg/h

(C) 1200 kg/h

(D) 1900 kg/h

PROBLEM 14

A lagooned bio-solid sludge contaminated with low concentrations of PCBs is currently being destroyed by incineration. Equipment is in place to dewater the sludge to 35% moisture and dry it to 5% moisture prior to incineration. Local community opposition to incineration has prompted the industry to investigate alternative destruction or disposal technologies.

14.1. What resources would likely be most fruitful for identifying appropriate alternative technologies to incineration?

   (A) alternative technology equipment vendor and trade publications

   (B) publications of the USEPA Office of Research and Development

   (C) publications of the USEPA Chemical Emergency Preparedness and Prevention Office

   (D) publications of the USEPA Superfund Innovative Technology Evaluation (SITE) Program

14.2. Which of the following is probably not an appropriate alternative technology to incineration for the lagooned bio-solid sludge?

   (A) thermal desorption and chemical dehalogenation

   (B) supercritical water oxidation

   (C) deep well injection

   (D) chemical destruction and fixation

14.3. Which of the following technologies would probably require drying the sludge to 5% moisture?

   (A) thermal desorption and chemical dehalogenation

   (B) supercritical water oxidation

   (C) chemical destruction and fixation

   (B) biological oxidation

14.4. What conditions are necessary for regulatory approval of a CAMU for disposing of the dewatered sludge on site?

   (A) When disposal of the waste at an approved landfill would require transporting the waste over public roadways for distances greater than 100 miles.

   (B) When the CAMU is constructed as part of a corrective action requirement.

   (C) When the facility will be abandoned after completing closure requirements.

   (D) When land disposal of the waste is selected as the disposal alternative.

PROBLEM 15

The following problems apply to the transportation of hazardous materials.

15.1. What regulations control the transportation of hazardous materials in the United States?

   (A) Title 40 CFR Parts 172–268

   (B) Title 29 CFR Part 1910

   (C) Title 49 CFR Parts 100–199

   (D) Title 40 CFR Part 300

15.2. Which of the following is not an objective of the regulations controlling transportation of hazardous materials?

   (A) defining storage requirements for materials awaiting transportation

   (B) defining packaging requirements for materials to be transported

   (C) classifying materials to be transported

   (D) communicating hazards of the materials to be transported to persons handling them during transport and to emergency response personnel

15.3. Which of the following is not included in the Hazardous Materials Table (HMT)?

   (A) proper shipping names

   (B) hazard class or division

   (C) chemical formula

   (D) identification numbers

15.4. What does the designation "ORM–D" represent?

   (A) oxidizer/reactive material

   (B) other regulated materials

   (C) ordinary recyclable materials

   (D) other registered material

15.5. What are Hazard Class I materials?

   (A) explosives

   (B) gases

   (C) flammable and combustible liquids

   (D) flammable solids

15.6. What form is required for shipping hazardous materials?

   (A) No specific form is required, but specific information must be included on whatever form is used.

   (B) the Uniform Hazardous Waste Manifest, but only if the material is hazardous waste

   (C) the DOT shipping form 172

   (D) both (B) and (C)

PROBLEM 16

The results of a time study and route analysis for curbside residential waste collection in a planned community are as follows.

population	600
per-capita solid waste generation rate	0.9 kg/person·d
number of residences	285
average driving time between residences	18 s
average pick-up/load time at each residence	45 s
travel time from truck yard to route start	38 min
travel time from route end to landfill	63 min
time to unload at landfill	20 min
travel time from landfill to truck yard	45 min
truck compacted waste capacity	12 m^3
truck compaction ratio	2.6:1
typical waste as-discarded density	140 kg/m^3

16.1. What is the daily total as-discarded volume of waste requiring collection?

- (A) 1.4 m^3/d
- (B) 3.9 m^3/d
- (C) 7.7 m^3/d
- (D) 29 m^3/d

16.2. What is the total time available to a single crew in one 8 h day for curbside collection?

- (A) 120 min
- (B) 230 min
- (C) 310 min
- (D) 480 min

16.3. What is the total compacted waste volume one crew can collect during one 8 h work day if collection occurs once weekly?

- (A) 7.2 m^3
- (B) 11 m^3
- (C) 15 m^3
- (D) 27 m^3

16.4. How many days should be scheduled for one crew to collect all the waste generated by the community?

- (A) 1 d
- (B) 2 d
- (C) 3 d
- (D) 4 d

PROBLEM 17

A city of 50,000 people generates municipal solid waste at a rate of 2 kg/person·d with an as-discarded density of 120 kg/m^3. The city has evaluated options for replacing its fleet of collection trucks and has selected trucks with a 23 m^3 compacted capacity. The trucks can compact the waste to 575 kg/m^3. Truck crews of three men each work 8 h/d, 5 d/wk, with collections occurring at each stop once weekly. A truck crew can fill a truck in 175 min. The transfer station is centrally located and can be reached in approximately 25 min from any collection route in the city. 15 min is required for unloading at the transfer station.

17.1. What is the total compacted volume of solid waste collected from the city each week?

- (A) 830 m^3/wk
- (B) 1200 m^3/wk
- (C) 5800 m^3/wk
- (D) 21 000 m^3/wk

17.2. How many trips can a single truck complete in one day?

- (A) 1
- (B) 2
- (C) 3
- (D) 4

17.3. How many trucks are required to meet the weekly collection schedule?

- (A) 4
- (B) 5
- (C) 6
- (D) 7

PROBLEM 18

The following problems apply to waste minimization.

18.1. Does a statutory requirement exist for the reduction or elimination of hazardous waste generation in the United States?

- (A) No, but USEPA has established a nonenforceable policy to encourage reduction of hazardous waste generation.
- (B) No, specific policy or regulatory authority does not exist for waste reduction; all efforts are strictly voluntary.
- (C) Yes, amendments to RCRA specifically declare that U.S. national policy is to reduce or eliminate hazardous waste generation.
- (D) Yes, SARA specifically declares that U.S. national policy is to reduce or eliminate hazardous waste generation.

18.2. Waste minimization can occur through which of the following activities?

(A) by improving plant operations such as house-keeping, equipment maintenance, and materials handling

(B) by substituting raw materials that produce fewer hazardous constituent by-products

(C) by redesigning processes to modify by-product characteristics

(D) by all of the above

18.3. Which alternative represents the highest priority of potential waste minimization activities?

(A) reuse and recovery

(B) eliminate generation

(C) recycle

(D) treatment

18.4. What is not a potential impediment to implementing waste minimization activities?

(A) economics

(B) public disclosure

(C) customer specifications for manufactured products

(D) unknown consequences of changing proven processes

PROBLEM 19

The following problems address household and small-generator waste management issues.

19.1. What does "a substance or material . . . that has been determined by the Secretary of Transportation to be capable of posing an unreasonable risk to health, safety, and property when transported in commerce" define?

(A) solid waste

(B) hazardous waste

(C) hazardous substance

(D) hazardous material

19.2. What criteria and conditions determine when a household waste is hazardous?

(A) The way a product containing a potentially hazardous material is used.

(B) when the waste satisfies any one of the four hazardous waste characteristics defined under RCRA in Title 40 CFR Part 261

(C) when the product is retained after its labeled expiration date

(D) when the waste has been stored for greater than 90 days

19.3. When alternative uses for waste are applied that do not require reprocessing or other significant modification, what is the result?

(A) reduction

(B) recycling

(C) reuse

(D) substitution

19.4. Which phrase most accurately reflects the meaning of a common labeling term?

(A) A "biodegradable" substance presents less hazard than does a "non-toxic" substance.

(B) A "combustible" substance will ignite at a lower temperature than will a "flammable" substance.

(C) "Danger" indicates a greater hazard than does "warning."

(D) "Caution" identifies a substance of greater toxicity than does "poison."

19.5. For older products, what is the best procedure for responding to an accidental release or a potentially harmful exposure?

(A) Follow label spill response or first-aid instructions.

(B) If the label is not intact or does not provide instructions, evacuate the area and dilute the release with copious amounts of water.

(C) Contact a poison control center or local hazardous materials response crew.

(D) Since older products deteriorate, take no action unless a negative impact is observable.

PROBLEM 20

The following problems relate to solid waste recycling and source reduction.

20.1. Which of the following is a primary impediment to widespread recycling in the United States?

(A) insufficient quantity of recyclable materials

(B) poorly developed and unstable market for recyclable materials

(C) low public and industry participation

(D) low economic and measurable environmental incentives

20.2. What is source reduction?

(A) restriction of raw material markets to reduce the manufacture of high waste-yielding products

(B) prohibition of recyclable materials in municipal solid waste

(C) design, manufacture, and use of products to reduce quantity and toxicity of waste produced at the end of a product's useful life

(D) development and application of alternative technologies to landfilling and incineration

20.3. Which of the following does not represent an example of successful source reduction?

(A) disposable cameras

(B) high mileage automobile tires

(C) concentrated fruit juices

(D) lightweight plastic milk containers

20.4. What sustainable recycling rate is reasonable in the United States for materials destined to become municipal solid waste?

(A) 10 to 20%

(B) 25 to 50%

(C) 60 to 75%

(D) 80 to 100%

PROBLEM 21

The following problems address emergency response and contingency planning.

21.1. Who is the target audience for SARA Title III?

(A) employers of contaminated site workers

(B) local emergency planning committees

(C) individual citizens

(D) hazardous waste generators

21.2. Who is the target audience for OSHA regulations in Title 29 CFR 1910.120?

(A) employers of contaminated site workers

(B) local emergency planning committees

(C) individual citizens

(D) hazardous waste generators

21.3. What is the primary focus of the Emergency Planning and Community Right-to-Know Act (EPCRA) of 1986?

(A) emergency planning and notification

(B) reporting, to include MSDS and chemical release and inventory forms

(C) both (A) and (B)

(D) neither (A) nor (B)

21.4. What is the function of the National Response Team (NRT)?

(A) to provide emergency response services at federal facilities including DOE and DOD sites

(B) to coordinate federal response and preparedness activities for hazardous material releases

(C) to enforce emergency response regulation compliance by state and local government agencies

(D) to regulate emergency response equipment manufacturers and emergency response service providers

21.5. What are the three basic steps of hazard analysis in emergency response planning?

(A) vulnerability analysis, risk analysis, response functions

(B) hazard identification, incident assessment, resource management

(C) hazard identification, vulnerability analysis, risk analysis

(D) incident assessment, response functions, resource management

PROBLEM 22

Soil contamination by petroleum hydrocarbons is widespread and a common objective of remedial action programs in the United States. The following problems pertain to the remediation of petroleum-contaminated soils.

22.1. Which of the following is not typically an important factor for *ex situ* bioremediation of petroleum contaminated soils?

(A) moisture content

(B) soil grain size distribution

(C) acclimated microbial population

(D) nutrient availability

22.2. Which of the following is not an appropriate *in situ* remedial alternative for heavier fuel oil-contaminated soils?

(A) thermal desorption and recovery

(B) ambient vapor extraction

(C) enhanced biodegradation

(D) containment

22.3. How is soil retention capacity for petroleum hydrocarbons affected by soil moisture?

(A) Retention capacity increases with increasing soil moisture.

(B) Retention capacity decreases with increasing soil moisture.

(C) Retention capacity increases as soil moisture approaches field capacity, then declines as soil moisture continues to increase.

(D) Retention capacity is not significantly influenced by soil moisture.

22.4. When might "no action" be acceptable as a remedial action alternative?

(A) when the contamination is widely dispersed, making remediation expensive

(B) when the contamination occurs in rural areas where property values are low

(C) when the contamination is confined and the contaminated soils are fine-grained

(D) when the contamination is volatile and the contaminated soils are coarse-grained

PROBLEM 23

A manufacturer has historically used a chromic acid-based solution to impregnate timbers and other structural wood products. The chromic acid solution was allowed to drain from the impregnated timbers onto bare soil thereby contaminating the underlying shallow unconfined aquifer. Soil samples were collected at the site, and a soil leaching test was conducted. The results of the leaching test are presented in the following table. The saturated soil porosity and soil bulk density are 0.34 and 1.8 g/cm^3, respectively.

test sample	Cr(VI) concentration in leachate (mg/L)	Cr(VI) adsorbed onto soil (mg/kg)
1	31	17
2	20	10
3	13	7.5
4	10	7.0
5	7.1	4.5

23.1. What is the soil distribution coefficient for hexavalent chromium?

(A) 0.50 mL/g

(B) 0.59 mL/g

(C) 0.63 mL/g

(D) 0.70 mL/g

23.2. What is the retardation factor of the hexavalent chromium in the soil/groundwater system?

(A) 0.5

(B) 1.5

(C) 3.0

(D) 4.0

23.3. Will a "pump and treat" remedial alternative be successful in quickly removing the hexavalent chromium from the soil/groundwater system?

(A) Yes, because the distribution coefficient is relatively small.

(B) Yes, because the distribution coefficient is relatively large.

(C) No, because the distribution coefficient is relatively small.

(D) No, because the distribution coefficient is relatively large.

23.4. Will infiltration through contaminated overlying soils be a continuing source of hexavalent chromium to the groundwater ?

(A) Yes, hexavalent chromium will leach slowly from the soil to the groundwater.

(B) Yes, hexavalent chromium will leach rapidly from the soil to the groundwater.

(C) No, hexavalent chromium will not leach from the soil.

(D) No, hexavalent chromium will be rapidly reduced to trivalent chromium.

PROBLEM 24

The following problems relate to radioactivity and radioactive wastes.

24.1. What is the atomic structure condition that causes radioactivity?

(A) an unstable neutron:proton ratio

(B) an unstable electron:proton ratio

(C) an unstable electron:neutron ratio

(D) a deficient number of electrons in the outer electron orbital

24.2. What radioactive particle is characterized by high energy and very small mass?

(A) alpha particles
(B) beta particles
(C) gamma particles
(D) delta particles

24.3. Which types of radioactive wastes are most commonly associated with non-defense industries, hospitals, and laboratories?

(A) high-level waste (HLW)
(B) low-level waste (LLW)
(C) transuranic waste (TRU)
(D) all of the above

24.4. What is the half-life of uranium-238?

(A) 3.05 min
(B) 3.83 d
(C) 1622 yr
(D) 4.51×10^9 yr

24.5. How many years are required for 99% of the original uranium-238 atoms in a radioactive waste to degrade to another isotope?

(A) 20 min
(B) 25 d
(C) 10,800 yr
(D) 3.0×10^{10} yr

PROBLEM 25

The following problems apply to the Resource Conservation and Recovery Act of 1976 (RCRA).

25.1. How is "solid waste" generally defined under RCRA?

(A) any discarded solid material
(B) any discarded liquid or solid material
(C) any discarded gaseous, liquid, or solid material
(D) any discarded solid material including containers from gases and liquids, but not gases and liquids

25.2. What are the four characteristics that RCRA applies to define hazardous waste?

(A) ignitable, non-biodegradable, corrosive, and volatile
(B) ignitable, corrosive, toxic, and reactive
(C) volatile, corrosive, toxic, and conductive
(D) volatile, non-biodegradable, toxic, and reactive

25.3. What is the RCRA program for addressing clean-up of contaminated sites?

(A) remedial action (RA)
(B) remedial response (RR)
(C) corrective remedial measures (CMR)
(D) corrective action (CA)

25.4. What types of facilities and/or activities require RCRA hazardous waste permits?

(A) facilities that generate hazardous wastes
(B) facilities that treat, store, or dispose hazardous wastes
(C) transporters of hazardous waste
(D) all of the above

25.5. Who qualifies as a "small quantity hazardous waste generator"?

(A) facilities that generate less than 100 kg in a calendar month
(B) facilities that generate less than 1000 kg during any 90 d period
(C) facilities that store hazardous waste for less than 90 d
(D) facilities that generate less than 1000 kg in a calendar month

PROBLEM 26

The following problems apply to the Comprehensive Environmental Response, Compensation, and Liability Act of 1980 (CERCLA), the Superfund Amendments and Reauthorization Act of 1986 (SARA), and the National Contingency Plan (NCP).

26.1. What issues are addressed by Title III of SARA?

(A) contaminated site ranking protocol
(B) contaminated site clean-up guidance
(C) compliance with applicable or relevant and appropriate requirements (ARARs)
(D) community right-to-know and emergency preparedness

26.2. How do "removal actions" differ from "remedial actions" under CERCLA?

(A) Remedial actions can be applied to both NPL and non-NPL sites, but removal actions apply to NPL sites only.

(B) Removal actions are typically long-term projects implemented to achieve a permanent remedy.

(C) Removal actions are subject to remedial investigation (RI), but not feasibility study (FS), activities prior to implementing response actions.

(D) Removal actions are short-term responses to stabilize or clean up sites that pose immediate threat to human health or the environment.

26.3. What is a Record of Decision (ROD)?

(A) the official report that documents the site background and describes the selected remedy

(B) the technical plans and specifications required to physically implement a selected remedy

(C) the listing of alternative remedial measures applicable to a site

(D) the documentation of the implemented remedy and ongoing operation and maintenance activities

26.4. Which of the following are not available to USEPA as Superfund enforcement tools?

(A) *de minimis* settlements

(B) TSD facility permit revocation

(C) mixed funding settlements

(D) nonbinding allocations of responsibility (NBARs)

26.5. Which of the following is a mechanism used by USEPA to support state activities authorized under CERCLA?

(A) alternative remedial contract strategy (ARCS)

(B) cooperative agreement (CA)

(C) administrative order on consent (AOC)

(D) technical assistance grant (TAG)

27.1. Under what USEPA Office does the Office of Environmental Justice operate?

(A) Office of Enforcement and Compliance

(B) Office of International Activities

(C) Office of Policy

(D) Office of Solid Waste and Emergency Response

27.2. How is environmental justice defined?

(A) equitable allocation of environmental resources among community, recreational, business, and industrial interests

(B) appropriate assessment of financial penalties to business and industry to allow full compensation for environmental damage

(C) fair and meaningful involvement of all people regardless of race or income with respect to development, implementation, and enforcement of environmental laws, regulations, and policies

(D) equitable reallocation of environmental resources to developing nations

27.3. Where are USEPA's environmental justice goals implemented?

(A) They are implemented in the environmental impact statement (EIS) and environmental assessment (EA) process.

(B) They are implemented in the CERCLA remedial investigation (RI) and feasibility study (FS) process.

(C) They are implemented in the RCRA permitting process.

(D) They are implemented in international technologies cooperation activities.

27.4. What are not appropriate environmental justice activities?

(A) identifying alternative sites where impacts to susceptible populations will be avoided

(B) adopting pollution-prevention practices to reduce impacts

(C) enforcing policies of involuntary relocation (IR) away from impacting sites

(D) reducing the size and intensity of impacting actions

PROBLEM 27

The following problems apply to environmental justice and the USEPA Office of Environmental Justice.

PROBLEM 28

An industrial facility generates waste from four different processes. The wastes are characterized as follows.

characteristic	waste 1	waste 2	waste 3	waste 4
ignitability (flash point, °C)	51	112	>200	64
corrosivity (pH)	4.3	11.6	1.7	8.4
reactivity (reactive, yes/no)	no	no	no	no
toxicity - contaminant concentration (mg/L)	vinyl chloride 105	–	–	–
chemical constituents	–	phenol	–	–

28.1. Which wastes are hazardous?

(A) wastes 1 and 2

(B) wastes 1 and 3

(C) wastes 2 and 3

(D) wastes 2 and 4

28.2. If the flash point of waste 4 were 60°C, would the hazardous waste characterization change?

(A) Yes, the waste would become hazardous.

(B) No, the waste would remain hazardous.

(C) Yes, the waste would become nonhazardous.

(D) No, the waste would remain nonhazardous.

28.3. If phenol were eliminated from waste 2, would the hazardous waste characterization change?

(A) Yes, the waste would become hazardous.

(B) No, the waste would remain hazardous.

(C) Yes, the waste would become nonhazardous.

(D) No, the waste would remain nonhazardous.

28.4. If waste 3 were treated by simple neutralization, would the hazardous waste characterization change?

(A) Yes, the waste would become hazardous.

(B) No, the waste would remain hazardous.

(C) Yes, the waste would become nonhazardous.

(D) No, the waste would remain nonhazardous.

28.5. If vinyl chloride were eliminated from waste 1, would the hazardous waste characterization change?

(A) Yes, the waste would become hazardous.

(B) No, the waste would remain hazardous.

(C) Yes, the waste would become nonhazardous.

(D) No, the waste would remain nonhazardous.

Solid Waste and Hazardous Waste Solutions

SOLUTION 1

Calculate the moisture content and the density based on a 100 kg sample of the waste.

component	discarded mass (kg)	discarded moisture (%)	dry mass (kg)	discarded dry density (kg/m³)	discarded dry volume (m³)	unit dry energy (kJ/kg)	total dry energy (kJ)
paper	44	6	41.0	85	0.48	16 750	686 750
garden	17	60	6.8	105	0.065	6500	44 200
food	11	70	3.3	290	0.011	4650	15 345
cardboard	9	5	8.6	50	0.17	16 300	140 180
wood	7	20	5.6	240	0.023	18 600	104 160
plastic	7	2	6.9	65	0.11	32 600	224 940
miscellaneous	5	8	4.6	480	0.010	7000	32 200
	100		76.8		0.87		1 247 775

$$\text{dry mass, kg} = (\text{discarded mass, kg})\left(\frac{1 - \%\ \text{moisture}}{100\%}\right)$$

$$\text{discarded dry volume, m}^3 = \frac{\text{dry mass, kg}}{\text{discarded dry density, kg/m}^3}$$

$$\text{total dry energy, kJ} = (\text{dry mass, kg})(\text{unit dry energy, kJ/kg})$$

1.1. The moisture content is

$$\frac{100\ \text{kg}}{100\ \text{kg}} - \frac{76.8\ \text{kg}}{100\ \text{kg}} = \boxed{23.2\ \text{kg}/100\ \text{kg}\quad (23\ \text{kg}/100\ \text{kg})}$$

The answer is A.

1.2. The dry waste bulk density is

$$\frac{76.8\ \text{kg}}{0.87\ \text{m}^3} = \boxed{88\ \text{kg/m}^3}$$

The answer is C.

1.3. The total waste generated daily is

$$(82{,}000\ \text{people})\left(2.7\ \frac{\text{kg}}{\text{person·d}}\right) = 221\,400\ \text{kg/d}$$

The dry waste energy content is

$$\left(\frac{1\,247\,775\ \text{kJ}}{100\ \text{kg}}\right)\left(221\,400\ \frac{\text{kg}}{\text{d}}\right)$$

$$= \boxed{2.8 \times 10^9\ \text{kJ/d}}$$

The answer is C.

1.4. c = specific heat of water
= 4.184 kJ/kg·°C
q_b = heat (energy) added to raise water to vaporization temperature, kJ
T_2 = vaporization temperature of water at 1 atm
= 100°C
T_1 = ambient temperature, assume = 20°C
m = mass of water, kg

$$m = \left(221\,400\ \frac{\text{kg}}{\text{d}}\right)\left(\frac{23\ \text{kg water}}{100\ \text{kg waste}}\right) = 50\,922\ \text{kg/d}$$

$$q_b = mc(T_2 - T_1)$$

$$= \left(50\,922\ \frac{\text{kg}}{\text{d}}\right)\left(4.184\ \frac{\text{kJ}}{\text{kg·°C}}\right)(100°\text{C} - 20°\text{C})$$

$$= 1.7 \times 10^7\ \text{kJ/d}$$

$$q_v = \text{heat of vaporization} = \left(2258\ \frac{\text{kJ}}{\text{kg}}\right)\left(50\,922\ \frac{\text{kg}}{\text{d}}\right)$$

$$= 1.15 \times 10^8\ \text{kJ/d}$$

$$q_T = q_b + q_v = 1.7 \times 10^7\ \frac{\text{kJ}}{\text{d}} + 1.15 \times 10^8\ \frac{\text{kJ}}{\text{d}}$$

$$\boxed{= 1.32 \times 10^8\ \text{kJ/d}}$$

The energy content of the waste at 2.8×10^9 kJ/d is greater than the energy required at 1.32×10^8 kJ/d to dry the waste. The energy content of the waste is adequate to dry the waste without adding energy from an external source.

The answer is A.

SOLUTION 2

Use a 100 kg sample as the basis for calculations.

(a)

waste component	dry mass (kg/100 kg)	elemental chemical mass (kg/100 kg)				
		C	H	O	N	S
food	4.9	2.4	0.31	1.8	0.13	0.020
glass/metal	3.2	–	–	–	–	–
paper	12.6	5.5	0.76	5.5	0.038	0.025
plastic	8.7	5.2	0.63	2.0	–	–
wood debris	2.1	1.0	0.13	0.90	0.0042	0.0021
yard clippings	29.5	14	1.8	11	1.0	0.089
	61	28	3.6	21	1.2	0.14

The moisture content is

$$100\ \text{kg} - 61\ \text{kg} = 39\ \text{kg}$$

The chemical content of moisture is

$$\text{hydrogen} = \left(\frac{2}{18}\right)(39\ \text{kg}) = 4.3\ \text{kg}$$

$$\text{oxygen} = \left(\frac{16}{18}\right)(39\ \text{kg}) = 34.7\ \text{kg}$$

The total hydrogen is

$$3.6\ \text{kg} + 4.3\ \text{kg} = 7.9\ \text{kg}$$

The total oxygen is

$$21\ \text{kg} + 34.7\ \text{kg} = 56\ \text{kg}$$

From table (a), the following table can be derived.

(b)

element	mass (kg)	mole weight (kg/kmol)	kmol	% mass
carbon	28	12	2.3	30
hydrogen	7.9	1	7.9	8.5
oxygen	56	16	3.5	60
nitrogen	1.2	14	0.086	1.3
sulfur	0.14	32	0.0044	0.15
	93			99.95

2.1. From kmol in table (b),

(c)

element	mole ratio S = 1
carbon	523
hydrogen	1795
oxygen	795
nitrogen	20
sulfur	1

The chemical formula is $\boxed{C_{523}H_{1795}O_{795}N_{20}S}$.

The answer is D.

2.2. From kmol in table (b),

(d)

element	mole ratio N = 1
carbon	27
hydrogen	92
oxygen	41
nitrogen	1

The chemical formula is $\boxed{C_{27}H_{92}O_{41}N}$.

The answer is C.

2.3. The energy content in kJ/kg is

$$337C + (1428)\left(H - \frac{O}{8}\right) + 95S$$

$$C, H, O, S = \text{elements, \% of total}$$

From % mass in table (b), the energy content is

$$(337)(30) + (1428)\left(8.5 - \frac{60}{8}\right) + (95)(0.15)$$

$$\boxed{= 11\,552\ \text{kJ/kg} \quad (12\,000\ \text{kJ/kg})}$$

The answer is B.

SOLUTION 3

3.1. The annual solid waste generated is

$$(25{,}000 \text{ people}) \left(1.85 \; \frac{\text{kg}}{\text{person·d}}\right) \left(365 \; \frac{\text{d}}{\text{yr}}\right)$$

$$= \boxed{1.7 \times 10^7 \text{ kg/yr}}$$

The answer is D.

3.2.

$$C_aH_bO_cN_d + (0.25)(4a - b - 2c + 3d)H_2O \longrightarrow$$
$$(0.125)(4a + b - 2c - 3d)CH_4$$
$$+ (0.125)(4a - b + 2c + 3d)CO_2 + dNH_3$$

a, b, c, d = moles of carbon, hydrogen, oxygen, and nitrogen, respectively

$C_{40}H_{86}O_{37}N$
$$+ (0.25)\big((4)(40) - 86 - (2)(37) + (3)(1)\big)H_2O \longrightarrow$$
$$(0.125)\big((4)(40) + 86 - (2)(37) - (3)(1)\big)CH_4$$
$$+ (0.125)\big((4)(40) - 86 + (2)(37) + (3)(1)\big)CO_2$$
$$+ 1NH_3$$
$$C_{40}H_{86}O_{37}N + 0.75H_2O \longrightarrow 21CH_4 + 19CO_2 + 1NH_3$$

1 mole of solid waste as $C_{40}H_{86}O_{37}N$ yields 21 moles of methane gas as CH_4.

The molecular weight of the waste is

$$(40)\left(12 \; \frac{\text{g}}{\text{mol}}\right) + (86)\left(1 \; \frac{\text{g}}{\text{mol}}\right)$$
$$+ (37)\left(16 \; \frac{\text{g}}{\text{mol}}\right) + (1)\left(14 \; \frac{\text{g}}{\text{mol}}\right)$$
$$= 1172 \text{ g/mol}$$

$$\frac{\left(1.7 \times 10^7 \; \dfrac{\text{kg waste}}{\text{yr}}\right) \left(21 \; \dfrac{\text{kmol methane}}{\text{kmol waste}}\right)}{1172 \; \dfrac{\text{kg waste}}{\text{kmol waste}}}$$
$$= 3.0 \times 10^5 \text{ kmol methane/yr}$$

Assume that methane behaves as an ideal gas so that $PV = nRT$ applies.

P = pressure = 1 atm
V = volume of methane, m^3
n = moles of methane = 3.0×10^5 kmol/yr
R = universal gas law constant
$\quad = 8.2 \times 10^{-5}$ m³·atm/mol·K
T = temperature = 298K

$$V = \frac{\left(\left(3.0 \times 10^5 \; \dfrac{\text{kmol}}{\text{yr}}\right) \times \left(8.2 \times 10^{-5} \; \dfrac{\text{m}^3 \text{·atm}}{\text{mol·K}}\right)(298\text{K})\right)}{(1 \text{ atm})\left(\dfrac{1 \text{ kmol}}{1000 \text{ mol}}\right)}$$

$$= \boxed{7.3 \times 10^6 \text{ m}^3/\text{yr}}$$

The answer is B.

3.3. The potential annual savings is

$$\left(7.3 \times 10^6 \; \frac{\text{m}^3}{\text{yr}}\right)\left(\frac{\$0.21}{\text{m}^3}\right)$$

$$= \boxed{\$1{,}533{,}000/\text{yr} \quad (\$1{,}500{,}000/\text{yr})}$$

The answer is B.

3.4. Landfill gas contains methane at about 60% by volume. To be used as a replacement fuel for natural gas, the landfill gas must be dried to remove water vapor and scrubbed to remove gases that are noncombustible.

The answer is C.

SOLUTION 4

4.1. D^* = apparent (effective) diffusion, m^2/s
$\quad \omega$ = tortuosity coefficient, unitless = 0.06
$\quad D_d$ = diffusion coefficient = 2.03×10^{-9} m^2/s

$$D^* = \omega D_d$$
$$= (0.06)\left(2.03 \times 10^{-9} \; \frac{\text{m}^2}{\text{s}}\right)$$
$$= \boxed{1.2 \times 10^{-10} \text{ m}^2/\text{s}}$$

The answer is C.

4.2. $\quad t$ = time, s
$\quad x$ = distance = 4.0 m
$\quad C_o$ = beginning concentration = 1106 mg/L
$\quad C$ = concentration at some future time, t
$\quad\quad = 10$ mg/L

$$\frac{C}{C_o} = \text{erfc}\left(\frac{x}{(2)(D^*t)^{0.5}}\right)$$

$$a = \left(\frac{x}{(2)(D^*t)^{0.5}}\right)$$

$$\text{erfc}(a) = \frac{10 \; \dfrac{\text{mg}}{\text{L}}}{1106 \; \dfrac{\text{mg}}{\text{L}}} = 0.00904$$

$$a = 1.83 = \left(\frac{x}{(2)(D^*t)^{0.5}} \right)$$

$$= \left(\frac{4.0 \text{ m}}{(2)\left(\left(1.2 \times 10^{-10} \frac{\text{m}^2}{\text{s}} \right) t \right)^{0.5}} \right)$$

$$t = 9.95 \times 10^9 \text{ s} = \boxed{316 \text{ yr} \quad (320 \text{ yr})}$$

The answer is C.

SOLUTION 5

5.1. The annual mass of waste landfilled is

$$(1 - 0.28)(125{,}000 \text{ people}) \left(1.6 \frac{\text{kg}}{\text{person·d}} \right) \left(365 \frac{\text{d}}{\text{yr}} \right)$$

$$= \boxed{5.3 \times 10^7 \text{ kg/yr}}$$

The answer is B.

5.2. The annual in-place volume of waste landfilled is

$$\frac{5.3 \times 10^7 \dfrac{\text{kg}}{\text{yr}}}{\left(850 \dfrac{\text{kg}}{\text{m}^3} \right) \left(10\,000 \dfrac{\text{m}^2}{\text{ha}} \right)} = \boxed{6.2 \text{ ha·m/yr}}$$

The answer is B.

5.3. The annual cover volume is

$$\frac{6.2 \dfrac{\text{ha·m}}{\text{yr}}}{5} = \boxed{1.24 \text{ ha·m/yr} \quad (1.2 \text{ ha·m/yr})}$$

The answer is C.

5.4. The total landfill volume is

$$\left(6.2 \frac{\text{ha·m}}{\text{yr}} + 1.2 \frac{\text{ha·m}}{\text{yr}} \right) (30 \text{ yr})$$

$$= \boxed{222 \text{ ha·m} \quad (220 \text{ ha·m})}$$

The answer is B.

SOLUTION 6

6.1.

category	criteria	WF	site A R	site A WR	site B R	site B WR	site C R	site C WR	site D R	site D WR
soil	permeability	1	4	4	3	3	5	5	2	2
	heterogeneities	3	2	6	2	6	1	3	3	9
geology	seismic activity	3	5	15	5	15	5	15	4	12
groundwater	quality/use	7	6	42	4	28	3	21	5	35
	gradient/depth	5	7	35	4	20	4	20	3	15
hydrology	topography	7	3	21	5	35	2	14	4	28
	streams/lakes	4	4	16	3	12	6	24	4	16
community	population	3	6	18	7	21	3	9	3	9
	land uses	7	7	49	7	49	2	14	5	35
				206		189		125		161
	relative rating			1.65		1.51		1.00		1.29

WR = weighted rating = (WF)(R)

Site C has the best rating.

The answer is C.

6.2. Landfills are prohibited within the borders of wetlands, critical habitat areas, and 100 yr flood plains, but not within U.S. military installations.

The answer is C.

6.3. The USDA Soil Conservation Service (SCS) provides descriptive soils information for most areas of the United States on a county-by-county basis.

The answer is D.

6.4. Although criteria of economics, technical feasibility, and environmental protection are important to landfill siting, the most critical criterion is gaining public acceptance.

The answer is B.

SOLUTION 7

7.1. $d_1, d_2 \ldots$ = thickness of each layer, cm
d = total thickness of all layers
$(d_1 + d_2 + \ldots)$, cm
$K_1, K_2 \ldots$ = hydraulic conductivity of each layer, cm/s
K = overall hydraulic conductivity for all layers, cm/s

$$K = \frac{d}{\dfrac{d_1}{K_1} + \dfrac{d_2}{K_2} + \dfrac{d_3}{K_3} + \dfrac{d_4}{K_4} + \dfrac{d_5}{K_5}}$$

$$= \frac{30 \text{ cm} + 15 \text{ cm} + 60 \text{ cm} + 3200 \text{ cm} + 30 \text{ cm}}{\left(\begin{array}{c} \dfrac{30 \text{ cm}}{10^{-3} \text{ cm·s}^{-1}} + \dfrac{15 \text{ cm}}{10^{-2} \text{ cm·s}^{-1}} \\[2mm] + \dfrac{60 \text{ cm}}{10^{-6} \text{ cm·s}^{-1}} + \dfrac{3200 \text{ cm}}{10^{-3} \text{ cm·s}^{-1}} \\[2mm] + \dfrac{30 \text{ cm}}{10^{-1} \text{ cm·s}^{-1}} \end{array}\right)}$$

$$= \boxed{5.3 \times 10^{-5} \text{ cm/s}}$$

The answer is B.

7.2. q = unit downward percolating leachate flow to each lateral, $\text{m}^3/\text{m·s}$
K = overall hydraulic conductivity
  $= 5.3 \times 10^{-5}$ cm/s
L_s = lateral spacing, m
L_l = lateral length = 150 m
N = number of laterals draining an area = 2

$$q = \frac{KL_sL_l}{NL_l}$$

$$= \frac{\left(\begin{array}{c} \left(5.3 \times 10^{-5} \dfrac{\text{cm}}{\text{s}}\right) L_s (150 \text{ m}) \\[2mm] \times \left(\dfrac{1 \text{ m}^3}{10^6 \text{ cm}^3}\right)\left(10^4 \dfrac{\text{cm}^2}{\text{m}^2}\right) \end{array}\right)}{(2)(150 \text{ m})}$$

$$= (2.6 \times 10^{-7} \text{ m}^3/\text{m}^2\text{·s})L_s$$

K_δ = hydraulic conductivity of the drainage layer
  $= 10^{-1}$ cm/s
d_o = maximum desired leachate head = 30 cm
d_c = distance of leachate lateral above the liner = 0

$$L_s = \frac{2K_d(d_o^2 - d_c^2)}{q}$$

$$= \frac{(2)\left(10^{-1} \dfrac{\text{cm}}{\text{s}}\right)\left(\begin{array}{c}(30)^2 \text{ cm}^2 \\ - (0)^2 \text{ cm}^2\end{array}\right)\left(\dfrac{1 \text{ m}^3}{10^6 \text{ cm}^3}\right)}{\left(2.6 \times 10^{-7} \dfrac{\text{m}^3}{\text{m}^2\text{·s}}\right) L_s}$$

$$L_s^2 = 692 \text{ m}^2$$

$$L_s = \boxed{26.3 \text{ m} \quad (26 \text{ m})}$$

The answer is B.

7.3. Q = flow rate to each lateral

$$q = \left(2.6 \times 10^{-7} \frac{\text{m}^3}{\text{m}^2\text{·s}}\right)(26 \text{ m}) = 6.8 \times 10^{-6} \text{ m}^3/\text{m·s}$$

$$Q = \left(6.8 \times 10^{-6} \frac{\text{m}^3}{\text{m·s}}\right)(150 \text{ m}) = \boxed{0.0010 \text{ m}^3/\text{s}}$$

The answer is C.

7.4. A_x = lateral cross-sectional area, m^2
  v_c = self-cleaning velocity in the lateral
    $= 0.6$ m/s

$$A_x = \frac{Q}{v_c} = \frac{0.0010 \dfrac{\text{m}^3}{\text{s}}}{0.6 \dfrac{\text{m}}{\text{s}}} = 0.0017 \text{ m}^2$$

The lateral diameter is

$$\left(\frac{4A_x}{\pi}\right)^{1/2} = \left(\frac{(4)(0.0017 \text{ m}^2)}{\pi}\right)^{1/2}$$

$$= \boxed{0.0465 \text{ m} \quad (4.7 \text{ cm})}$$

The answer is B.

SOLUTION 8

Subtitle C landfill liner requirements are defined in Title 40 CFR Section 264.301. The liner requirements are illustrated in the following figure.

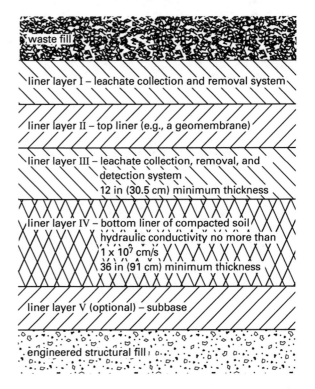

8.1. From the figure, liner layer III represents the leachate collection, detection, and removal system (LCDRS).

The answer is C.

8.2. From the figure, liner layer II represents the geomembrane.

The answer is B.

8.3. From the figure, liner layer IV represents the soil liner.

The answer is D.

8.4. From the figure, liner layer I represents the leachate collection and removal system (LCRS).

The answer is A.

SOLUTION 9

9.1. Gas venting for a landfill that accepted hazardous waste would have to maintain the integrity of the cap against infiltration of surface runoff. Surface-sealed gravel cell or trench vents, vertical vapor wells, and vented perforated pipes would all satisfy this criterion. However, surface-exposed gravel cell or trench vents would not.

The answer is A.

9.2. The likely composition of the gas produced by the landfill, assuming it to be typical of other landfills, is 40% CO_2, 60% CH_4, and less than 1% others.

The answer is D.

9.3. C_a = average methane production for arid regions, $m^3/min \cdot 10^9$ kg

W = landfilled waste from which methane is drawn
$\quad = 3.0 \times 10^8$ kg

$C_a = 5.87W$

$$= \frac{\left(\left(5.87 \, \frac{m^3}{min} \right) (3.0 \times 10^8 \text{ kg}) \times \left(1440 \, \frac{min}{d} \right) \left(365 \, \frac{d}{yr} \right) \right)}{10^9 \text{ kg}}$$

$$= \boxed{9.3 \times 10^5 \text{ m}^3/\text{yr}}$$

The answer is C.

9.4. $t_1 = 0$ yr
$t_2 = 15$ yr
C_1 = methane generation rate at time t_1
$\quad = 9.3 \times 10^5 \text{ m}^3/\text{yr}$
C_2 = methane generation rate at time t_2, m^3/yr
ζ = decay constant (assumed) = 0.05 yr^{-1}

$$\frac{C_2}{C_1} = \exp\left(-\zeta(t_2 - t_1) \right)$$

$$C_2 = \left(9.3 \times 10^5 \, \frac{m^3}{yr} \right) \exp \left(\begin{array}{c} -(0.05 \text{ yr}^{-1}) \\ \times (15 \text{ yr} - 0 \text{ yr}) \end{array} \right)$$

$$= \boxed{4.4 \times 10^5 \text{ m}^3/\text{yr}}$$

The answer is C.

SOLUTION 10

10.1.

$$\frac{\left(\begin{array}{c} (0.32) \left(300\,000 \, \frac{kg}{d} \right) \left(\frac{1}{12} \right) \\ + (\text{daily sludge mass}) \left(\frac{1}{90} \right) \end{array} \right)}{(0.32) \left(300\,000 \, \frac{kg}{d} \right) + \text{daily sludge mass}} = 1/30$$

Solving for daily sludge mass gives

$$\boxed{216\,000 \text{ kg/d} \quad (220\,000 \text{ kg/d})}$$

The answer is C.

10.2. Assume the moisture content of the compostable solid waste is 38%.

$$\frac{\left(\begin{array}{c} (0.32)(0.38) \left(300\,000 \, \frac{kg}{d} \right) \\ + (1 - 0.18) \left(220\,000 \, \frac{kg}{d} \right) \end{array} \right)}{(0.32) \left(300\,000 \, \frac{kg}{d} \right) + 220\,000 \, \frac{kg}{d}} = 0.686$$

The moisture content of the mixture is $\boxed{69\%}$.

The answer is D.

10.3. The total compostable mixture of waste and sludge is

$$(0.32) \left(300\,000 \, \frac{kg}{d} \right) + 220\,000 \, \frac{kg}{d} = 316\,000 \text{ kg/d}$$

Remove $69\% - 60\% = 9\%$ moisture.

$$\left(316\,000 \ \frac{\text{kg}}{\text{d}}\right)(0.09) = \boxed{28\,440 \text{ kg/d} \quad (28\,000 \text{ kg/d})}$$

The answer is D.

10.4. The thermophilic microbes that are responsible for the biological processes occurring in the compost pile thrive at temperatures above 50°C. As the temperature approaches 60°C, pathogens that may be present in the compost begin to die. Consequently, optimum temperatures for composting are generally in the range of 55°C to 60°C.

The answer is D.

10.5. Composting is a thermophylic aerobic process.

The answer is C.

SOLUTION 11

Injection wells are regulated under the Safe Drinking Water Act through programs primarily administered by individual state regulatory agencies.

11.1. Class I injection wells are used for injection of hazardous wastes.

The answer is A.

11.2. Class V injection wells are used for aquifer recharge.

The answer is D.

11.3. Class II injection wells are typically associated with oil and gas production dry holes.

The answer is B.

11.4. Groundwater recharge wells are not the only wells used for injection into a groundwater formation. Class IV injection wells are those that are used to inject hazardous waste into or above a formation that has an underground source of drinking water within one-quarter mile of the bore hole.

The answer is C.

SOLUTION 12

12.1. C_{CO_2} = emission CO_2 concentration
   $= 1600$ kg/h
   C_{CO} = emission CO concentration
   $= 0.13$ kg/h

The combustion efficiency is

$$\left(\frac{C_{CO_2}}{C_{CO_2} + C_{CO}}\right)(100\%)$$

$$= \left(\frac{1600 \ \frac{\text{kg}}{\text{h}}}{1600 \ \frac{\text{kg}}{\text{h}} + 0.13 \ \frac{\text{kg}}{\text{h}}}\right)(100\%)$$

$$= \boxed{99.9919\%}$$

The answer is C.

12.2. W_{in} = waste mass feed rate in
   W_{out} = waste mass feed rate out
   $= 0.0433$ kg/h

$$W_{in} = \left(\frac{27\%}{100\%}\right)\left(1800 \ \frac{\text{kg}}{\text{h}}\right) = 486 \text{ kg/h}$$

The destruction and removal efficiency is

$$\text{DRE} = \left(\frac{W_{in} - W_{out}}{W_{in}}\right)(100\%)$$

$$= \left(\frac{486 \ \frac{\text{kg}}{\text{h}} - 0.0433 \ \frac{\text{kg}}{\text{h}}}{486 \ \frac{\text{kg}}{\text{h}}}\right)(100\%)$$

$$= \boxed{99.9911\%}$$

The answer is B.

12.3. P_C = particulate concentration corrected for oxygen in the stack gas, mg/m^3
   P_M = measured concentration of particulate
   $= 203$ mg/m^3
   Y = measured O_2 concentration in the stack gas
   $= 4\%$

$$P_C = \frac{14P_M}{21 - Y} = \frac{(14)\left(203 \ \frac{\text{mg}}{\text{m}^3}\right)}{21 - 4}$$

$$= \boxed{167 \text{ mg/m}^3 \quad (170 \text{ mg/m}^3)}$$

The answer is C.

SOLUTION 13

13.1. The waste feed rate is

$$\left(1200\ \frac{L}{h}\right)(0.95)\left(1\ \frac{kg}{L}\right) = \boxed{1140\ kg/h}$$

The answer is B.

13.2. The mass flow of CO_2 produced is

$$(0.80)\left(1140\ \frac{kg}{h}\right)(0.78C)\left(\frac{44\ g\ CO_2}{12\ g\ C}\right)$$

$$= \boxed{2608\ kg/h \quad (2600\ kg/h)}$$

The mass flow of HCl produced is

$$(0.80)\left(1140\ \frac{kg}{h}\right)(0.02Cl)\left(\frac{36.5\ g\ HCl}{35.5\ g\ Cl}\right)$$

$$= \boxed{18.8\ kg/h \quad (19\ kg/h)}$$

The answer is C.

13.3. The stoichiometric mass flow of O_2 required for combustion of carbon is

$$(0.80)\left(1140\ \frac{kg}{h}\right)(0.78C)\left(\frac{32\ g\ O_2}{12\ g\ C}\right)$$

$$= \boxed{1897\ kg/h \quad (1900\ kg/h)}$$

The answer is D.

SOLUTION 14

14.1. Of the resources listed, publications of the USEPA Superfund Innovative Technology Evaluation (SITE) Program would be most fruitful for identifying appropriate alternative technologies to incineration. Publications of the USEPA Office of Research and Development and the USEPA Chemical Emergency Preparedness and Prevention Office would include limited, if any, information regarding alternative technologies. Alternative technology equipment vendor publications and trade publications typically do not provide objective information appropriate for evaluating alternative technologies.

The answer is D.

14.2. Although their degree of development to full-scale applications varies, technologies such as thermal desorption and chemical dehalogenation, supercritical water oxidation, and chemical destruction and fixation are potentially appropriate alternative technologies to incineration for the bio-solid sludge in the lagoon. Deep-well injection would likely not be acceptable either as a technical alternative or to local community activists opposing incineration.

The answer is C.

14.3. Thermal desorption requires heating the sludge to temperatures that will volatilize the target contaminants. High moisture in the sludge will limit the ability to heat the sludge to temperatures high enough for volatilization to occur. Therefore, for thermal desorption, drying the sludge to 5% moisture would be required for effective application of the technology. Supercritical water oxidation, chemical destruction and fixation, and biological oxidation can all be applied to sludges with relatively high moisture content.

The answer is A.

14.4. Regulatory approval of a CAMU for disposing of the dewatered sludge on-site would only occur if the CAMU were constructed as part of a required corrective action. Transporting the waste over public roadways for distances greater than 100 mi, abandoning a facility after completing closure requirements, and applying land disposal as the disposal alternative are not justifications for regulatory approval of a CAMU.

The answer is B.

SOLUTION 15

15.1. Although other regulations, such as Title 40 CFR Part 263, address transportation of hazardous wastes and materials, these regulations reference the U.S. Department of Transportation regulations in Title 49 Parts 100 to 199. The Title 49 regulations are known as the Hazardous Materials Regulations and provide for the comprehensive regulation of hazardous waste, hazardous substance, and hazardous materials transportation in the United States.

The answer is C.

15.2. Regulations controlling transportation of hazardous materials define packaging requirements for materials to be transported, classify materials to be transported, and provide information communicating the hazards of the materials to be transported to persons handling them during transport and to emergency response personnel. The hazardous materials regulations do not define storage requirements for materials awaiting transportation.

The answer is A.

15.3. Among other information, the Hazardous Materials Table (Title 49 CFR Section 172.101) provides proper shipping names, hazard class or division, and identification numbers for hazardous materials, but not chemical formulas.

The answer is C.

15.4. The designation "ORM-D" represents other regulated materials (Title 49 CFR Section 173.144).

The answer is B.

15.5. Hazard Class 1 materials, as defined in Title 49 CFR Section 173.50, are explosives.

The answer is A.

15.6. In general, no specific form is required for shipping hazardous materials, but specific information must be included on whatever form is used. However, if hazardous wastes are being transported, the shipping papers must include the Uniform Hazardous Waste Manifest.

The answer is A.

SOLUTION 16

16.1. The total daily as-discarded waste volume requiring collection is

$$\frac{(600 \text{ people})\left(0.9 \frac{\text{kg}}{\text{person·d}}\right)}{140 \frac{\text{kg}}{\text{m}^3}} = \boxed{3.9 \text{ m}^3/\text{d}}$$

The answer is B.

16.2. The available collection time is the total time minus the noncollection task time.

$$(8 \text{ h})\left(60 \frac{\text{min}}{\text{h}}\right) - \left(\begin{array}{c} 38 \text{ min} + 63 \text{ min} \\ + 20 \text{ min} \\ + 45 \text{ min} \end{array}\right)$$

$$= \boxed{314 \text{ min} \quad (310 \text{ min})}$$

The answer is C.

16.3. The possible stops per day are

$$\frac{\text{total time}}{\dfrac{\text{time}}{\text{stop}}} + 1 = \frac{(310 \text{ min})\left(60 \frac{\text{s}}{\text{min}}\right)}{18 \text{ s} + 45 \text{ s}} + 1 = 296$$

The number of residences is 285.

Enough time is available to collect at all residences in a single day since 285 is less than 296.

Since waste can be collected during a single 8 h day and collection occurs once weekly, the total compacted volume collected is

$$\frac{\left(0.9 \frac{\text{kg}}{\text{person·d}}\right)(600 \text{ people})\left(7 \frac{\text{d}}{\text{wk}}\right)}{\left(140 \frac{\text{kg}}{\text{m}^3}\right)\left(\frac{2.6 \text{ m}^3}{1 \text{ m}^3}\right)}$$

$$= \boxed{10.4 \text{ m}^3/\text{wk} \quad (11 \text{ m}^3/\text{wk})}$$

The answer is B.

16.4. From Prob. 16.3, time is available to collect all waste in one day, and the truck capacity of 12 m³ is adequate since it is greater than the daily compacted volume collected of 11 m³.

The answer is A.

SOLUTION 17

17.1. The weekly compacted volume collected is

$$\frac{(50{,}000 \text{ people})\left(2 \frac{\text{kg}}{\text{person·d}}\right)\left(7 \frac{\text{d}}{\text{wk}}\right)}{575 \frac{\text{kg}}{\text{m}^3}}$$

$$= \boxed{1217 \text{ m}^3/\text{wk} \quad (1200 \text{ m}^3/\text{wk})}$$

The answer is B.

17.2. The time available for collection is

$$\left(8 \frac{\text{h}}{\text{d}}\right)\left(60 \frac{\text{min}}{\text{h}}\right) = 480 \text{ min}$$

task	task time (min)	cumulative time (min)
leave	25	25
collect 1	175	200
return	25	225
unload	15	240
leave	25	265
collect 2	175	440
return	25	465
unload	15	480

A single truck can complete two trips in one day.

The answer is B.

17.3.
$$\dfrac{\left(1200\ \frac{m^3}{wk}\right)\left(\frac{1\ wk}{5\ d}\right)\left(\frac{1\ d}{2\ trips}\right)}{23\ \frac{m^3}{truck\cdot trip}} = 5.2\ trucks$$

Six trucks are required.

The answer is C.

SOLUTION 18

18.1. RCRA and amendments as codified in U.S. Code Title 42, Chapter 82, Section 6902 identifies reduction and elimination of hazardous waste as an objective of the statute. Section 6902(b) states "the Congress hereby declares it to be national policy of the United States that, wherever feasible, the generation of hazardous waste is to be reduced or eliminated as expeditiously as possible."

The answer is C.

18.2. Waste minimization can occur through improving plant operations such as housekeeping, equipment maintenance, and materials handling, through substituting raw materials that produce fewer hazardous constituent by-products, and through redesigning processes to modify by-product characteristics.

The answer is D.

18.3. The hierarchy of waste minimization is (1) eliminate generation, (2) reduce generation, (3) recycle, (4) reuse and recovery, (5) treatment, and (6) disposal.

The answer is B.

18.4. Waste minimization activities are potentially impeded by economics, customer specifications for manufactured products, and unknown consequence of changing proven processes. Public disclosure is typically not an issue associated with waste minimization activities.

The answer is B.

SOLUTION 19

19.1. A hazardous material is defined in Title 49 CFR Section 171.8 as "a substance or material... that has been determined by the Secretary of Transportation to be capable of posing an unreasonable risk to health, safety, and property when transported in commerce."

The answer is D.

19.2. Criteria and conditions that determine when a household waste is hazardous are determined by the way the product containing a potentially hazardous material is used.

The answer is A.

19.3. When alternative uses for a waste are applied that do not require reprocessing or other significant modification, reuse is the result.

The answer is C.

19.4. A biodegradable substance is not necessarily nontoxic and may or may not present a lesser hazard than a nontoxic substance. A combustible substance will not ignite at a lower temperature than will a flammable substance; "flammable" signifies a greater hazard than does "combustible." Likewise, "poison" indicates a greater hazard than does "caution." However, "danger" does indicate a greater hazard than does "warning."

The answer is C.

19.5. Because proper responses to an accidental release or potentially harmful exposure are continually being evaluated and improved, labeling information on older products may be obsolete. The best response for releases or exposures involving older products, therefore, is to contact a poison control center or local hazardous materials response crew.

The answer is C.

SOLUTION 20

20.1. A primary impediment to widespread recycling in the United States is a poorly developed and unstable market for recyclable materials. A sufficient quantity of recyclable materials and measurable economic and environmental incentives do exist. Public and industry participation is mandated in many jurisdictions.

The answer is B.

20.2. Source reduction is the design, manufacture, and use of products to reduce the quantity and toxicity of waste produced at the end of a product's useful life.

The answer is C.

20.3. High-mileage automobile tires, concentrated fruit juices, and lightweight plastic milk containers are all examples of successful source reduction, but disposable cameras are not.

The answer is A.

20.4. The most widely accepted reasonably sustainable recycling rate in the United States for materials destined to become municipal solid waste is 25% to 50%.

The answer is B.

SOLUTION 21

21.1. The target audience of SARA Title III is local emergency planning committees, including emergency response agencies such as police and fire departments.

The answer is B.

21.2. The target audience of OSHA regulations in Title 29 CFR 1910.120 are employers of contaminated site workers.

The answer is A.

21.3. The primary focus of the Emergency Planning and Community Right-to-Know Act (EPCRA) of 1986 is emergency planning and notification and reporting, to include MSDS and chemical release and inventory forms.

The answer is C.

21.4. The function of the National Response Team (NRT) is to coordinate federal response and preparedness activities for hazardous material releases.

The answer is B.

21.5. The three basic steps of hazard analysis in emergency response planning are (1) hazard identification, (2) vulnerability analysis, and (3) risk analysis.

The answer is C.

SOLUTION 22

22.1. Moisture content, an acclimated microbial population, and nutrient availability are all important factors for *ex situ* bioremediation of petroleum contaminated soils. Soil grain size distribution is much less important.

The answer is B.

22.2. Heavier fuel oil-contaminated soils may be appropriately remediated by *in situ* alternatives such as thermal desorption and recovery, enhanced biodegradation, and containment. Ambient vapor extraction would not be effective for heavier fuel oils since they are low in volatile fractions.

The answer is B.

22.3. Soil retention capacity for petroleum hydrocarbons decreases with increasing soil moisture.

The answer is B.

22.4. "No action" may be acceptable as a remedial action alternative when the contamination is confined and the contaminated soils are fine-grained.

The answer is C.

SOLUTION 23

23.1.

test sample	Cr(VI) concentration in leachate (mg/L)	Cr(VI) adsorbed onto soil (mg/kg)	distribution coefficient (mL/g)
1	31	17	0.55
2	20	10	0.50
3	13	7.5	0.58
4	10	7.0	0.70
5	7.1	4.5	0.63

The distribution coefficient for each test sample is the Cr(VI) adsorbed onto soil divided by the Cr(VI) concentration in leachate.

The distribution coefficient is

$$\frac{\left(\begin{array}{c} 0.55 \frac{mL}{g} + 0.50 \frac{mL}{g} + 0.58 \frac{mL}{g} \\ + 0.70 \frac{mL}{g} + 0.63 \frac{mL}{g} \end{array}\right)}{5}$$

$$= \boxed{0.59 \text{ mL/g}}$$

The answer is B.

23.2. R = retardation factor, unitless
ρ_b = soil density = 1.8 g/cm^3
θ = saturated soil porosity = 0.34
K_d = distribution coefficient = 0.59 mL/g

$$R = 1 + \left(\frac{\rho_b}{\theta}\right) K_d$$

$$= 1 + \frac{\left(1.8 \ \frac{g}{cm^3}\right)\left(0.59 \ \frac{mL}{g}\right)\left(\frac{1 \ cm^3}{1 \ mL}\right)}{0.34}$$

$$= \boxed{4.12 \quad (4.0)}$$

The answer is D.

23.3. A "pump and treat" remedial alternative will probably not be successful in quickly removing the hexavalent chromium from the soil/groundwater system because the distribution coefficient is relatively large.

The answer is D.

23.4. Infiltration through contaminated overlying soils will be a continuing source of hexavalent chromium to the groundwater because the relatively large distribution coefficient will allow the hexavalent chromium to leach slowly from the soil.

The answer is A.

SOLUTION 24

24.1. The atomic structure that causes radioactivity is an unstable neutron:proton ratio.

The answer is A.

24.2. Alpha particles are relatively low energy particles that can be stopped by thick sheets of paper. Beta particles have higher energy than alpha particles but can be stopped by the equivalent of thin aluminum sheeting. Gamma particles, however, are characterized by high energy and very small mass and can only be stopped by lead sheeting or thick concrete.

The answer is C.

24.3. Low-level radioactive wastes (LLW) are most commonly associated with non-defense industries, hospitals, and laboratories. High-level (HLW) and transuranic (TRU) wastes are associated with nuclear fuel and weapons applications.

The answer is B.

24.4. The half-life of uranium-238 is 4.51×10^9 yr.

The answer is D.

24.5. Radioactive decay follows first-order kinetics.

$t_{1/2} = $ half-life $= 4.51 \times 10^9$ yr
$\quad k = $ first-order rate constant, yr^{-1}

$$t_{1/2} = \frac{\ln(0.5)}{k}$$

$$k = \frac{\ln(0.5)}{4.51 \times 10^9 \ \text{yr}} = -1.54 \times 10^{-10} \ \text{yr}^{-1}$$

$\quad t = $ time, yr
$C_o = $ initial concentration
$\quad C = $ concentration at 99% degradation

$$C = C_o(1 - 0.99) = 0.01 C_o$$

$$t = \frac{\ln\left(\dfrac{C}{C_o}\right)}{k} = \frac{\ln\left(\dfrac{0.01 \ C_o}{C_o}\right)}{-1.54 \times 10^{-10} \ \text{yr}^{-1}}$$

$$= \boxed{3.0 \times 10^{10} \ \text{yr}}$$

The answer is D.

SOLUTION 25

25.1. In Title 40 CFR 261.2, RCRA defines solid waste as "any discarded material" whether solid, liquid, or gas.

The answer is C.

25.2. In Title 40 CFR Part 261, Subpart C, RCRA defines hazardous waste according to characteristics of ignitability, corrosivity, toxicity, and reactivity.

The answer is B.

25.3. The RCRA program for clean up of contaminated sites is corrective action (CA) as defined in Title 40 CFR 264.100.

The answer is D.

25.4. According to Title 40 CFR 264, facilities that treat, store, or dispose of hazardous waste must be permitted. Although hazardous waste generators and transporters are required to obtain an EPA identification number, they are generally not permitted.

The answer is B.

25.5. According to Title 40 CFR 260.10, a small quantity generator is "a generator who generates less than 1000 kg of hazardous waste in a calendar month."

The answer is D.

SOLUTION 26

26.1. SARA Title III addresses issues pertaining to community right-to-know and emergency preparedness.

The answer is D.

26.2. Removal actions are short-term responses to stabilize or clean up sites that pose immediate threats to human health or the environment. Remedial actions are longer-term projects designed to provide a permanent remedy.

The answer is D.

26.3. A record of decision (ROD) is the official report that documents the site background and describes the remedy selected.

The answer is A.

26.4. *De minimis* settlements, mixed fund settlements, and nonbinding allocation of responsibility are all Superfund enforcement tools available to the USEPA, but revocation of TSD facility permits is not.

The answer is B.

26.5. Cooperative agreements are used to transfer funds from the USEPA to states. Alternative remedial contract strategy, administrative order on consent, and technical assistant grant are not mechanisms used to support state activities.

The answer is B.

SOLUTION 27

27.1. The Office of Environmental Justice operates under the USEPA Office of Enforcement and Compliance.

The answer is A.

27.2. Environmental justice is the "fair and meaningful involvement of all people regardless of race or income with respect to development, implementation, and enforcement of environmental laws, regulations, and policies."

The answer is C.

27.3. The USEPA's policy is to implement goals that promote environmental justice in the environmental impact statement (EIS) and environmental assessment (EA) process.

The answer is A.

27.4. Appropriate environmental justice activities include identifying alternative sites that will not impact susceptible populations, adopting pollution prevention practices to reduce impacts, and reducing the size and intensity of impacting actions. Enforcing policies of involuntary relocation (IR) away from impacting sites is not an appropriate environmental justice activity.

The answer is C.

SOLUTION 28

28.1. Waste 1 is hazardous because of ignitability (flash point less than 60°C) according to Title 40 CFR 261.21 and toxicity (vinyl chloride concentration greater than 0.2 mg/L) according to Title 40 CFR 24. Waste 3 is hazardous because of corrosivity (pH less than 2) according to Title 40 CFR 261.22.

The answer is B.

28.2. No. The characteristic of ignitability defined in Title 40 CFR 261.21 requires the flash point to be less than, not less than or equal to, 60°C for the waste to be hazardous.

The answer is D.

28.3. No. Waste 2 is not hazardous with phenol included (phenol is not listed in Title 40 CRF 261.33). Removing phenol from the waste would not change its characterization as nonhazardous waste.

The answer is D.

28.4. Yes. Simple neutralization would increase the pH to above 2. According to Title 40 CFR 261.22, a waste is hazardous by corrosivity if its pH is less than or equal to 2 or greater than or equal to 12.5.

The answer is C.

28.5. No. If vinyl chloride were removed from the waste, the waste would still be hazardous because of the ignitability characteristic.

The answer is B.

Section IV
Environmental Assessments, Remediation, and Emergency Response

- Environmental Assessments

- Remediation

- Public Health and Safety

Environmental Assessments

PROBLEM 1

Three alternative systems for hazardous materials storage and handling are characterized by the reliability ratings presented below. Components occur in process sequence from A to B to C.

system	component	type	reliability
1	A	series	0.93
	B	series	0.97
	C	series	0.96
2	A	series	0.98
	B-1	parallel	0.83
	B-2	parallel	0.81
	C	series	0.98
3	A	series	0.95
	B-1	parallel	0.78
	B-2	parallel	0.80
	B-3	parallel	0.79

1.1. What is the overall reliability of system 1?

(A) 0.87

(B) 0.90

(C) 0.93

(D) 0.95

1.2. What is the overall reliability of system 2?

(A) 0.83

(B) 0.86

(C) 0.89

(D) 0.93

1.3. If the reliabilities for system 3 are based on 1000 h of operation, what is the failure rate for system 3?

(A) 6.1×10^{-5} failures/h

(B) 9.2×10^{-5} failures/h

(C) 1.1×10^{-4} failures/h

(D) 1.8×10^{-4} failures/h

PROBLEM 2

Malignant brain tumors were discovered in two long-time employees who work in the Chemical Research Center (CRC) building at Apex Chemicals & Pharmaceuticals. The CRC employs 320 scientists and engineers and their support staff. Apex employs 1100 other workers at the same complex where the CRC is located. Only one of these other workers has developed similar cancers to the CRC employees although all workers share similar lifestyle habits.

2.1. What is the relative risk?

(A) 0.14

(B) 0.50

(C) 3.5

(D) 6.9

2.2. What is the attributable risk?

(A) 0.0015

(B) 0.0053

(C) 0.29

(D) 0.5

2.3. What is the odds ratio?

(A) 2.0

(B) 3.5

(C) 3.8

(D) 6.9

2.4. What is the relationship between risk and exposure from working in the CRC?

(A) A risk likely exists from exposure in the CRC.

(B) A risk likely does not exist from exposure in the CRC.

(C) The data are contradictory, but a risk likely does not exist.

(D) The data are inconclusive and no conclusions are possible.

PROBLEM 3

The following questions address definitions associated with human health and environmental risk assessment.

3.1. In risk assessment, to what does hazard refer?

(A) the identification of a risk

(B) the quantification of a risk

(C) the existence of a toxin

(D) the occurrence of exposure

3.2. What are the four basic elements of risk assessment?

(A) hazard identification, population characterization, chemical assessment, risk characterization

(B) hazard identification, dose-response assessment, exposure assessment, risk characterization

(C) dose-response assessment, chemical assessment, population characterization, risk characterization

(D) dose-response assessment, chemical assessment, exposure assessment, risk characterization

3.3. How does the maximum exposed individual (MEI) differ from the reasonably maximally exposed (RME) individual?

(A) The RME has a longer life span than does the MEI.

(B) The MEI eats only locally produced food; the RME does not.

(C) The MEI is receptor specific; the RME is not.

(D) The RME exposure occurs at the same physical location; the MEI exposure does not.

3.4. What is the weight-of-evidence category for "probable human carcinogen"?

(A) Group A

(B) Group B

(C) Group C

(D) Group D

3.5. For noncarcinogens, which among the following represents the highest dose?

(A) threshold dose

(B) reference dose (RfD)

(C) dose equal to the no observed adverse effects level (NOAEL)

(D) dose equal to the lowest observed adverse effects level (LOAEL)

PROBLEM 4

A population of 25 000 is exposed over a period of 23 yr to the following VOCs in their drinking water supply.

trichloroethene	120 μg/L
1,1-dichloroethene	80 μg/L
methylene chloride	50 μg/L

4.1. What is the incremental lifetime cancer risk to the average adult from using the water supply?

(A) 5.4×10^{-8}

(B) 4.7×10^{-4}

(C) 1.4×10^{-3}

(D) 3.5×10^{-2}

4.2. Is the incremental lifetime cancer risk from drinking the contaminated water considered acceptable by commonly applied criteria?

(A) Yes, because the risk is less than 1 in 1 million.

(B) Yes, because the risk is greater than 1 in 1 million.

(C) No, because the risk is less than 1 in 1 million.

(D) No, because the risk is greater than 1 in 1 million.

4.3. In any given year, how many additional cancers would be expected to result among the exposed population?

(A) 1.9×10^{-5} cancers/yr

(B) 0.17 cancers/yr

(C) 35 cancers/yr

(D) 1600 cancers/yr

PROBLEM 5

Low levels of dioxins are emitted from a hazardous waste incinerator that results in annual average ambient air concentrations, measured as 2,3,7,8-TCDD, of 0.2 pg/m^3 in the adjacent community of 80 000 residents. The slope factor for 2,3,7,8-TCDD is 1.5×10^5 (mg/kg·d)$^{-1}$.

5.1. What is the incremental lifetime risk from the dioxin exposure to the average adult residing in the community for 35 yr?

(A) 1.2×10^{-7}

(B) 2.5×10^{-7}

(C) 4.4×10^{-6}

(D) 8.7×10^{-6}

5.2. What is the incremental lifetime risk from the dioxin exposure to a child living in the community from birth to age 6 yr?

(A) 3.6×10^{-9}

(B) 2.2×10^{-8}

(C) 8.6×10^{-7}

(D) 2.1×10^{-6}

5.3. If a 1 in 1 million or less risk is considered acceptable, what is the maximum permissible annual average ambient air concentration of dioxin as 2,3,7,8-TCDD for an adult exposed over a 35 yr period?

(A) 0.0034 pg/m^3

(B) 0.018 pg/m^3

(C) 0.047 pg/m^3

(D) 0.16 pg/m^3

PROBLEM 6

Toxicity values and concentrations are presented below for four chemicals found in a groundwater sample.

chemical	concentration (μg/L)	slope factor, SF (mg/L)$^{-1}$	reference dose, RfD (mg/L)
1,1-DCE	173	1.7×10^{-2}	0.315
MeCl	207	2.1×10^{-4}	2.1
PCE	879	1.5×10^{-3}	0.35
1,1,2-TCA	764	1.6×10^{-3}	0.14

6.1. What is the total carcinogenic risk factor from exposure to the chemicals?

(A) 0.0054

(B) 0.10

(C) 1.3

(D) 8.6

6.2. Which chemicals could be eliminated from further evaluation of carcinogenic risk?

(A) none of the chemicals

(B) MeCl

(C) 1,1-DCE and MeCl

(D) 1,1-DCE, MeCl, and PCE

6.3. What is the total noncarcinogenic risk factor from exposure to the chemicals?

(A) 0.0054

(B) 0.10

(C) 1.3

(D) 8.6

6.4. Which chemicals could be eliminated from further evaluation of noncarcinogenic risk?

(A) none of the chemicals

(B) MeCl

(C) 1,1-DCE and MeCl

(D) 1,1-DCE, MeCl, and PCE

PROBLEM 7

Two projects are being evaluated for implementation at an industrial facility. Project 1 requires upgrading chemical storage and conveyance equipment at a cost of $3 million. Project 2 requires developing and maintaining a hazardous waste reduction program at a cost of $2.6 million. The budget will only cover the cost of one of the projects. Project characteristics are defined as follows.

Project 1:

risk component	cost of current risk	cost of risk at project completion
A	2×10^6	2×10^4
B	6×10^5	6×10^4
C	9×10^5	4.5×10^3

Project 2:

risk component	cost of current risk	cost of risk at project completion
A	1.8×10^6	1.8×10^4
B	1×10^6	1×10^4
C	3×10^5	3×10^4

7.1. What is the risk reduction potentially realized by implementation of each project?

(A) 8.50×10^4, 5.80×10^4

(B) 1.14×10^6, 1.01×10^6

(C) 3.42×10^6, 3.04×10^6

(D) 3.50×10^6, 3.10×10^6

7.2. What is the risk reduction to cost ratio for each project?

(A) 0.028, 0.022

(B) 0.38, 0.39

(C) 1.14, 1.17

(D) 1.17, 1.19

7.3. Which project should be funded first?

(A) Project 1 because it has the smallest risk reduction to cost ratio.

(B) Project 1 because it has the greatest risk reduction to cost ratio.

(C) Project 2 because it has the smallest risk reduction to cost ratio.

(D) Project 2 because it has the greatest risk reduction to cost ratio.

PROBLEM 8

A workplace exposure to Co-60 through ingestion of drinking water is characterized as follows.

concentration = 1.6×10^{-11} Ci/L

ingestion rate = 1.4 L/d

exposure frequency = 200 d/yr

exposure duration = 30 yr

whole-body committed effective dose-equivalent per unit intake = 9.5×10^6 mrem/Ci

slope factor = 1.5×10^{-11} pCi^{-1}

Assume acceptable risk is 1 in 1 million and acceptable occupational exposure is 5 rem.

8.1. What is the intake of Co-60 through ingestion of the drinking water?

(A) 1.3×10^{-7} Ci

(B) 1.6×10^{-7} Ci

(C) 3.1×10^{-7} Ci

(D) 5.6×10^{-7} Ci

8.2. What is the whole-body committed effective dose-equivalent from the exposure?

(A) 1.2 mrem

(B) 1.5 mrem

(C) 2.9 mrem

(D) 5.3 mrem

8.3. What is the total lifetime cancer risk from the exposure?

(A) 0.93×10^{-6}

(B) 2.0×10^{-6}

(C) 4.4×10^{-6}

(D) 7.9×10^{-6}

8.4. Is the exposure level acceptable?

(A) Yes, the risk is less than 1 in 1 million and the whole-body dose is less than 5 rem.

(B) No, the risk is less than 1 in 1 million but the whole-body dose is greater than 5 rem.

(C) No, the risk is greater than 1 in 1 million even though the whole-body dose is less then 5 rem.

(D) No, the risk is greater than 1 in 1 million and the whole body dose is greater than 5 rem.

Environmental Assessments Solutions

1.1. R_A, R_B, R_C = reliability of components A, B, C

$$R_{system} = \text{overall system reliability}$$
$$= R_A R_B R_C$$
$$= (0.93)(0.97)(0.96)$$
$$= \boxed{0.87}$$

The answer is A.

1.2.
$$R_B = 1 - (1 - R_{B\text{-}1})(1 - R_{B\text{-}2})$$
$$= 1 - (1 - 0.83)(1 - 0.81)$$
$$= 0.97$$
$$R_{system} = (0.98)(0.97)(0.98)$$
$$= \boxed{0.93}$$

The answer is D.

1.3. R_t = reliability at time, t
λ = failure rate

$$R_t = e^{-\lambda t}$$

For component A,

$$0.95 = e^{-\lambda(1000)}$$
$$\lambda = 5.1 \times 10^{-5} \text{ failures/h}$$

For component B,

$$R_B = 1 - (1 - 0.78)(1 - 0.80)(1 - 0.79) = 0.99$$
$$0.99 = e^{-\lambda(1000)}$$
$$\lambda = 1.0 \times 10^{-5} \text{ failures/h}$$

The system failure rate is

$$5.1 \times 10^{-5} \frac{\text{failures}}{\text{h}} + 1.0 \times 10^{-5} \frac{\text{failures}}{\text{h}}$$
$$= \boxed{6.1 \times 10^{-5} \text{ failures/h}}$$

The answer is A.

	population with tumors	population without tumors
population exposed	$a = 2$	$b = 318$
population not exposed	$c = 1$	$d = 1099$

2.1. The relative risk is

$$\frac{a(c + d)}{c(a + b)} = \frac{(2)(1 + 1099)}{(1)(2 + 318)} = \boxed{6.9}$$

The answer is D.

2.2. The attributable risk is

$$\frac{a}{a + b} - \frac{c}{c + d} = \frac{2}{2 + 318} - \frac{1}{1 + 1099} = \boxed{0.0053}$$

The answer is B.

2.3. The odds ratio is

$$\frac{ad}{bc} = \frac{(2)(1099)}{(1)(318)} = \boxed{6.9}$$

The answer is D.

2.4. For relative risk and for odds ratio, the data suggests that as values increase above 1.0, the risk of contracting cancer from the exposure increases. The attributable risk is the difference between the odds of developing tumors with and without exposure. Values greater than 0.0 suggest increasing risk. The data suggest that a risk exists from the exposure in the CRC.

The answer is A.

3.1. In risk assessment, the term "hazard" refers to the existence of a toxin, independent of exposure. Risk cannot be identified or quantified unless exposure to a toxin occurs.

The answer is C.

3.2. The four basic elements of risk assessment are hazard identification, dose-response assessment, exposure assessment, and risk characterization.

The answer is B.

3.3. The maximum exposed individual (MEI) differs from the reasonably maximally exposed (RME) individual because, among other things, the MEI eats only locally produced food while the RME eats only 10% locally produced food.

The answer is B.

3.4. The weight-of-evidence category for "probable human carcinogen" is Group B. Group B chemicals are divided into one of the two subgroups, B1 or B2, depending on the type of evidence suggesting carcinogenicity. The Group A category is "human carcinogen," Group C is "possible human carcinogen," and Group D is "not classified."

The answer is B.

3.5. In increasing order, the four doses listed for non-carcinogens are: (1) reference dose (RfD), (2) dose equal to the no observed adverse effects level (NOAEL), (3) threshold dose, and (4) dose equal to the lowest observed adverse effects level (LOAEL).

The answer is D.

SOLUTION 4

4.1. Slope factors for VOCs detected in the groundwater are presented as follows.

chemical	concentration (μg/L)	slope factor, SF (mg/kg·d)$^{-1}$
trichloroethene (TCE)	120	1.1×10^{-2}
1,1-dichloroethene (1,1-DCE)	80	6.0×10^{-1}
methylene chloride (MeCl)	50	7.5×10^{-3}

USEPA residential exposure factors from ingestion of drinking water for adults are 2 L for daily intake and 70 kg body weight. The average lifetime is 70 yr.

CDI = chronic daily intake, mg/kg·d
$\quad C$ = concentration, mg/L
$\quad DI$ = daily intake = 2 L/d
$\quad \%A$ = % absorbed (assume 100%) = 1
$\quad ED$ = exposed duration = 23 yr
$\quad \%T$ = % of time exposed (assume 100%) = 1
$\quad BW$ = body weight = 70 kg
$\quad LT$ = lifetime = 70 yr

$$CDI = \frac{C\big((DI)(\%A)\big)\big((ED)(\%T)\big)}{(BW)(LT)}$$

$$TCE\ CDI = \frac{\left(\begin{array}{c}\left(120\ \frac{\mu g}{L}\right)\left(\frac{1\ mg}{1000\ \mu g}\right) \\ \times\left(\left(2\ \frac{L}{d}\right)(1)\right)\big((23\ yr)(1)\big)\end{array}\right)}{(70\ kg)(70\ yr)}$$

$$= 0.00113\ mg/kg{\cdot}d$$

$$1,1\text{-}DCE\ CDI = \frac{\left(\begin{array}{c}\left(80\ \frac{\mu g}{L}\right)\left(\frac{1\ mg}{1000\ \mu g}\right) \\ \times\left(\left(2\ \frac{L}{d}\right)(1)\right)\big((23\ yr)(1)\big)\end{array}\right)}{(70\ kg)(70\ yr)}$$

$$= 0.00075\ mg/kg{\cdot}d$$

$$MeCl\ CDI = \frac{\left(\begin{array}{c}\left(50\ \frac{\mu g}{L}\right)\left(\frac{1\ mg}{1000\ \mu g}\right) \\ \times\left(\left(2\ \frac{L}{d}\right)(1)\right)\big((23\ yr)(1)\big)\end{array}\right)}{(70\ kg)(70\ yr)}$$

$$= 0.00047\ mg/kg{\cdot}d$$

$$risk = (CDI)(SF)$$

$$risk\ TCE = \left(0.00113\ \frac{mg}{kg{\cdot}d}\right)$$
$$\times\left(1.1 \times 10^{-2}\left(\frac{mg}{kg{\cdot}d}\right)^{-1}\right)$$
$$= 1.2 \times 10^{-5}$$

$$risk\ 1,1\text{-}DCE = \left(0.00075\ \frac{mg}{kg{\cdot}d}\right)$$
$$\times\left(6.0 \times 10^{-1}\left(\frac{mg}{kg{\cdot}d}\right)^{-1}\right)$$
$$= 4.5 \times 10^{-4}$$

$$risk\ MeCl = \left(0.00047\ \frac{mg}{kg{\cdot}d}\right)$$
$$\times\left(7.5 \times 10^{-3}\left(\frac{mg}{kg{\cdot}d}\right)^{-1}\right)$$
$$= 3.5 \times 10^{-6}$$

$$total\ risk = 1.2 \times 10^{-5} + 4.5 \times 10^{-4} + 3.5 \times 10^{-6}$$
$$= \boxed{4.7 \times 10^{-4}}$$

The answer is B.

4.2. An incremental lifetime cancer risk of 1.0×10^{-6} (1 in 1 million) or less from exposure to a toxic chemical is usually considered acceptable. The combination of TCE, 1,1-DCE, and MeCl in the water produces a risk of 4.7×10^{-4} (1 in approximately 2200), which is greater than 1.0×10^{-6} and therefore considered unacceptable.

The answer is D.

4.3. The number of cancers contracted per year from drinking water is

$$\frac{\text{(risk)(population)}}{\text{LT}}$$

$$= \frac{\left(4.7 \times 10^{-4} \dfrac{\text{cancers}}{\text{person}}\right)(25{,}000 \text{ people})}{70 \text{ yr}}$$

$$= \boxed{0.17 \text{ cancers/yr}}$$

The answer is B.

SOLUTION 5

5.1. CDI = chronic daily intake, mg/kg·d
 C = concentration = 0.2 pg/m^3
 DI = daily intake = 20 m^3/d
 $\%A$ = % absorbed as fraction (assume 1)
 ED = exposed duration = 35 yr
 $\%T$ = % of time exposed as fraction (assume 1)
 BW = body weight = 70 kg
 LT = lifetime = 70 yr
 SF = slope factor = 1.5×10^5 (mg/kg·d)$^{-1}$

$$\text{CDI} = \frac{C\big((\text{DI})(\%A)\big)\big((\text{ED})(\%T)\big)}{(\text{BW})(\text{LT})}$$

$$= \frac{\left(0.2 \dfrac{\text{pg}}{\text{m}^3}\right)\left(\left(20 \dfrac{\text{m}^3}{\text{d}}\right)(1)\right)\big((35 \text{ yr})(1)\big)}{(70 \text{ kg})(70 \text{ yr})}$$

$$= 0.029 \text{ pg/kg·d}$$

$$\text{risk} = (\text{CDI})(\text{SF})$$

$$= \left(0.029 \dfrac{\text{pg}}{\text{kg·d}}\right)\left(1.5 \times 10^5 \left(\dfrac{\text{mg}}{\text{kg·d}}\right)^{-1}\right)$$

$$\times \left(10^{-9} \dfrac{\text{mg}}{\text{pg}}\right)$$

$$= \boxed{4.4 \times 10^{-6}}$$

The answer is C.

5.2. DI = 5 m^3/d for a child, and ED = 6 yr.

$$\text{CDI} = \frac{\left(0.2 \dfrac{\text{pg}}{\text{m}^3}\right)\left(\left(5 \dfrac{\text{m}^3}{\text{d}}\right)(1)\right)\big((6 \text{ yr})(1)\big)}{(15 \text{ kg})(70 \text{ yr})}$$

$$= 0.0057 \text{ pg/kg·d}$$

$$\text{risk} = (\text{CDI})(\text{SF})$$

$$= \left(0.0057 \dfrac{\text{pg}}{\text{kg·d}}\right)\left(1.5 \times 10^5 \left(\dfrac{\text{mg}}{\text{kg·d}}\right)^{-1}\right)$$

$$\times \left(10^{-9} \dfrac{\text{mg}}{\text{pg}}\right)$$

$$= \boxed{8.6 \times 10^{-7}}$$

The answer is C.

5.3. $\text{risk} = (\text{CDI})(\text{SF})$

$$= \left(\frac{C\big((\text{DI})(\%A)\big)\big((\text{ED})(\%T)\big)}{(\text{BW})(\text{LT})}\right)(\text{SF})$$

$$C = \frac{(\text{risk})(\text{BW})(\text{LT})}{\big((\text{DI})(\%A)\big)\big((\text{ED})(\%T)\big)(\text{SF})}$$

$$= \frac{(10^{-6})(70 \text{ kg})(70 \text{ yr})\left(10^9 \dfrac{\text{pg}}{\text{mg}}\right)}{\left(\begin{array}{c}\left(\left(20 \dfrac{\text{m}^3}{\text{d}}\right)(1)\right)\big((35 \text{ yr})(1)\big) \\ \times \left(1.5 \times 10^5 \left(\dfrac{\text{mg}}{\text{kg·d}}\right)^{-1}\right)\end{array}\right)}$$

$$= \boxed{0.047 \text{ pg/m}^3}$$

The answer is C.

SOLUTION 6

Carcinogenic risk calculations that apply to Probs. 6.1 and 6.2 are summarized in the following table.

chemical	concentration, C (μg/L)	slope factor, SF (mg/L)$^{-1}$	carcinogenic risk factor, R	relative risk, R_r
1,1-DCE	173	1.7×10^{-2}	0.0029	0.53
MeCl	207	2.1×10^{-4}	0.000043	0.0080
PCE	879	1.5×10^{-3}	0.0013	0.24
1,1,2-TCA	764	1.6×10^{-3}	0.0012	0.22
			0.0054	

$$R = (C \times 10^{-3})(\text{SF})$$

$$R_r = \frac{R}{\Sigma R}$$

6.1. The total carcinogenic risk factor is

$$0.0029 + 0.000043 + 0.0013 + 0.0012 = \boxed{0.0054}$$

The answer is A.

6.2. Typically, chemicals that present a relative risk of 0.01 (1% of total risk) or less are eliminated from further evaluation of carcinogenic risk. Based on this criterion, MeCl is the only chemical that could be eliminated from further study.

The answer is B.

Noncarcinogenic risk calculations that apply to Probs. 6.3 and 6.4 are summarized as follows.

chemical	concentration (μg/L)	reference dose, RfD (mg/L)	risk factor, R	relative risk, R_r
1,1-DCE	173	0.315	0.55	0.064
MeCl	207	2.1	0.099	0.012
PCE	879	0.35	2.5	0.29
1,1,2-TCA	764	0.14	5.5	0.64
			8.6	

$$R = \frac{C \times 10^{-3}}{\text{RfD}}$$

$$R_r = \frac{R}{\Sigma R}$$

6.3. The total noncarcinogenic risk factor is

$$0.55 + 0.099 + 2.5 + 5.5 = \boxed{8.6}$$

The answer is D.

6.4. Typically, chemicals that present a relative risk of 0.01 (1% of total risk) or less are eliminated from further evaluation of noncarcinogenic risk. Based on this criterion, none of the chemicals could be eliminated from further study.

The answer is A.

SOLUTION 7

7.1. R_{red} = cost benefit from total risk reduction
R_{current} = cost of each risk component under current conditions
R_{final} = cost of each risk component after project completion

$$R_{\text{red}} = \Sigma R_{\text{current}} - \Sigma R_{\text{final}}$$

For project 1,

$$R_{\text{red}} = (\$2 \times 10^6 + \$6 \times 10^5 + \$9 \times 10^5)$$
$$- (\$2 \times 10^4 + \$6 \times 10^4 + \$4.5 \times 10^3)$$
$$= \boxed{\$3.42 \times 10^6}$$

For project 2,

$$R_{\text{red}} = (\$1.8 \times 10^6 + \$1 \times 10^6 + \$3 \times 10^5)$$
$$- (\$1.8 \times 10^4 + \$1 \times 10^4 + \$3 \times 10^4)$$
$$= \boxed{\$3.04 \times 10^6}$$

The answer is C.

7.2. For project 1,

$$\frac{R_{\text{red}}}{\text{project cost}} = \frac{\$3.42 \times 10^6}{\$3.0 \times 10^6} = \boxed{1.14}$$

For project 2,

$$\frac{R_{\text{red}}}{\text{project cost}} = \frac{\$3.04 \times 10^6}{\$2.6 \times 10^6} = \boxed{1.17}$$

The answer is C.

7.3. The project with the greatest risk reduction to cost ratio should be funded first since it provides a greater benefit per unit of cost. Project 2 has the greatest risk reduction to cost ratio.

The answer is D.

SOLUTION 8

8.1. I = intake, Ci
C = concentration = 1.6×10^{-11} Ci/L
R_I = ingestion rate = 1.4 L/d
f_E = exposure frequency = 200 d/yr
D_t = exposure duration = 30 yr

$$I = CR_I f_E D_t$$
$$= \left(1.6 \times 10^{-11} \frac{\text{Ci}}{\text{L}}\right) \left(1.4 \frac{\text{L}}{\text{d}}\right) \left(200 \frac{\text{d}}{\text{yr}}\right) (30 \text{ yr})$$
$$= \boxed{1.34 \times 10^{-7} \text{ Ci} \quad (1.3 \times 10^{-7} \text{ Ci})}$$

The answer is A.

8.2. $H_{E,50}$ = whole body committed effective dose-equivalent, mrem

$h_{E,50}$ = whole body committed effective dose-equivalent per unit intake, 9.5×10^6 mrem/Ci

$$H_{E,50} = I h_{E,50}$$
$$= (1.3 \times 10^{-7}\text{ Ci}) \left(9.5 \times 10^6\ \frac{\text{mrem}}{\text{Ci}} \right)$$
$$= \boxed{1.24\text{ mrem} \quad (1.2\text{ mrem})}$$

The answer is A.

8.3. SF = slope factor = 1.5×10^{-11} pCi^{-1}

$$\text{risk} = (\text{SF})I$$
$$= (1.5 \times 10^{-11}\text{ pCi}^{-1})(1.3 \times 10^{-7}\text{ Ci}) \left(10^{12}\ \frac{\text{pCi}}{\text{Ci}} \right)$$
$$= \boxed{1.95 \times 10^{-6} \quad (2.0 \times 10^{-6})}$$

The answer is B.

8.4. No. The exposure level is not acceptable because the risk of 2 in 1 million is greater than the acceptable risk of 1 in 1 million even though the whole-body dose of 1.2 mrem is less than the acceptable occupational exposure of 5 rem.

The answer is C.

Remediation

PROBLEM 1

Subchronic animal studies were performed to evaluate the noncarcinogenic toxic effects of three chemical compounds. The chemicals were administered to the animals by ingestion of food with a 20% absorption efficiency. Results are shown in the following table.

chemical	NOAEL (mg/kg·d)	intake (mg/kg·d)
1	435	0.020
2	287	0.0092
3	329	0.014

If administered by ingestion of drinking water, the absorption efficiency is estimated to increase to 90%. Assume a safety factor of 5 for uncertainty based on professional judgment.

1.1. What is the administered oral reference dose for each chemical?

- (A) 0.0087 mg/kg·d, 0.0057 mg/kg·d, 0.0066 mg/kg·d
- (B) 0.044 mg/kg·d, 0.029 mg/kg·d, 0.033 mg/kg·d
- (C) 0.087 mg/kg·d, 0.057 mg/kg·d, 0.066 mg/kg·d
- (D) 0.17 mg/kg·d, 0.11 mg/kg·d, 0.13 mg/kg·d

1.2. What is the absorbed oral reference dose for each chemical through ingestion of food?

- (A) 0.0088 mg/kg·d, 0.0058 mg/kg·d, 0.0066 mg/kg·d
- (B) 0.017 mg/kg·d, 0.011 mg/kg·d, 0.013 mg/kg·d
- (C) 0.034 mg/kg·d, 0.022 mg/kg·d, 0.026 mg/kg·d
- (D) 0.44 mg/kg·d, 0.29 mg/kg·d, 0.33 mg/kg·d

1.3. What would be the equivalent administered oral reference dose for each chemical through ingestion of drinking water?

- (A) 0.015 mg/kg·d, 0.0099 mg/kg·d, 0.012 mg/kg·d
- (B) 0.019 mg/kg·d, 0.012 mg/kg·d, 0.014 mg/kg·d
- (C) 0.078 mg/kg·d, 0.051 mg/kg·d, 0.060 mg/kg·d
- (D) 0.096 mg/kg·d, 0.063 mg/kg·d, 0.073 mg/kg·d

1.4. Based on the hazard index, which chemicals should be targeted for remediation if the exposure route is ingestion of drinking water?

- (A) Chemicals 1 and 2 because their hazard index is less than 1.0.
- (B) Chemical 2 because its hazard index is greater than 1.0.
- (C) Chemicals 1 and 3 because their hazard index is greater than 1.0.
- (D) Chemicals 2 and 3 because their hazard index is less than 1.0.

PROBLEM 2

96 h toxicity tests using rainbow trout and conducted over a range of six concentrations for each of two chemical herbicides produced the following results.

herbicide 1		herbicide 2	
concentration (μg/L)	survival (%)	concentration (μg/L)	survival (%)
1000	100	200	94
1200	88	220	82
1400	73	240	69
1600	58	260	53
1800	46	280	44
2000	34	300	29

Tissue concentrations for rainbow trout exposed to the herbicides are as follows.

	exposure concentration (μg/L)	tissue concentration (μg/kg)
herbicide 1	0.9	38
herbicide 2	0.1	64

2.1. What is the lethal concentration for 50% of the test species after a 96 h exposure (96 h LC_{50}) in μg/L for each herbicide?

- (A) 1685 μg/L, 254 μg/L
- (B) 1706 μg/L, 273 μg/L
- (C) 1733 μg/L, 267 μg/L
- (D) 1837 μg/L, 271 μg/L

2.2. Based on 96 h LC_{50}, which herbicide is more toxic?

(A) herbicide 1, because it has a lower LC_{50} than herbicide 2

(B) herbicide 1, because it has a higher LC_{50} than herbicide 2

(C) herbicide 2, because it has a lower LC_{50} than herbicide 1

(D) herbicide 2, because it has a higher LC_{50} than herbicide 1

2.3. What is the bioconcentration factor for each herbicide?

(A) 0.024, 0.0016

(B) 34, 6.4

(C) 42, 640

(D) 240, 16

PROBLEM 3

Rural drinking water supplies in some western United States communities draw from groundwater that contains naturally occurring arsenic and fluoride at average concentrations of 109 μg/L and 1.7 mg/L, respectively. The groundwater is also used for watering vegetable gardens.

Residents of these communities tend to live out their lives within a few miles of the homes of their grandparents, parents, and siblings. Commodities from home food production, including canned home-grown fruits and vegetables, make up a large part of their diet.

Groundwater conditions and the toxicity characteristics of arsenic and fluoride are as follows.

For groundwater,

dissolved oxygen	0.8 mg/L
pH	8.0
temperature	16°C
specific conductivity	860 μS/cm
TDS	650 mg/L

For arsenic,

relative specie toxicity	As(V) less toxic than As(III)
MCL	50 μg/L
slope factor	5×10^{-5} $(\mu$g/L$)^{-1}$
oral R_fD	1×10^{-3} mg/kg·d
bioconcentration factor	44 mL/g

For fluoride,

MCL	4 mg/L
oral R_fD	6×10^{-2} mg/kg·d

3.1. Is the arsenic likely to be present in the drinking water as less toxic arsenate (As(V)) or as more toxic arsenite (As(III))?

(A) arsenate, because the groundwater dissolved oxygen concentration suggests oxidizing conditions

(B) arsenate, because the groundwater dissolved oxygen concentration suggests reducing conditions

(C) arsenite, because the groundwater dissolved oxygen concentration suggests oxidizing conditions

(D) arsenite, because the groundwater dissolved oxygen concentration suggests reducing conditions

3.2. Are the residents likely to be exposed to a significant cancer risk from ingesting the arsenic in their drinking water?

(A) No, the cancer risk is less than 1 in 10^6.

(B) No, the cancer risk is greater than 1 in 10^6.

(C) Yes, the cancer risk is less than 1 in 10^6.

(D) Yes, the cancer risk is greater than 1 in 10^6.

3.3. Are the residents likely to be exposed to significant noncarcinogenic health risks from ingesting the arsenic in their drinking water?

(A) No, the hazard index (HI) is less than 1.0.

(B) No, the hazard index (HI) is greater than 1.0.

(C) Yes, the hazard index (HI) is less than 1.0.

(D) Yes, the hazard index (HI) is greater than 1.0.

3.4. Is bioaccumulation of the arsenic in the residents' bodies over a lifetime of exposure likely to occur?

(A) Unknown; the bioconcentration factor applies to fish only and no conclusions can be made from the information provided regarding humans.

(B) No; although developed for exposure through ingestion of fish, the bioconcentration factor is low, suggesting unlikely bioaccumulation in humans through ingestion of water and vegetables.

(C) Yes, the bioconcentration factor suggests the potential for bioaccumulation and significant exposure exists through ingestion of water and vegetables.

(D) Yes, bioaccumulation always results from exposure to inorganic contaminants in drinking water and the food supply.

3.5. Are significant negative health consequences to the residents likely to result from ingesting the fluoride with the drinking water?

(A) No, the hazard index (HI) is less than 1.0 and the concentration is below the MCL.

(B) No, the hazard index (HI) is greater than 1.0 and the concentration is below the MCL.

(C) Yes, the hazard index (HI) is less than 1.0 and the concentration exceeds common fluoridation levels.

(D) Yes, the hazard index (HI) is greater than 1.0 and the concentration exceeds common fluoridation levels.

PROBLEM 4

The degradation sequence for perchloroethene to dichloroethene isomers is as follows.

perchloroethene (CCl_2-CCl_2)
⇓
trichloroethene ($CHCl$-CCl_2)
⇓
dichloroethene isomers
  (CH_2-CCl_2, *cis*- and *trans*-$CHCl$-$CHCl$)

Soil-water partition coefficients, Henry's constants, and solubility for selected chlorinated organic solvents are summarized in the following table.

compound	K_{oc} (mL/g)	K_H (atm·m³/mol)	solubility (mg/L)
chloroethane	42	0.0085	5740
chloroform	34	0.0038	9300
1,1-dichloroethene	217	0.021	2730
cis-1,2-dichloroethene	34	0.0037	3500
trans-1,2-dichloroethene	39	0.38	6300
methylene chloride	25	0.0032	16 700
perchloroethene	303	0.018	150
trichloroethene	152	0.010	1080
vinyl chloride	8400	2.8	1100

4.1. What is not a significant pathway for degradation of perchloroethene in a soil-groundwater system?

(A) biological oxidation

(B) chemical reduction

(C) deamination

(D) hydrolysis

4.2. What compound would follow the dichloroethene isomers in the degradation sequence for perchloroethene?

(A) chloroethane (CH_3-CH_2Cl)

(B) chloroform ($CHCl_3$)

(C) methylene chloride (CH_2Cl_2)

(D) vinyl chloride (CH_2-$CHCl$)

4.3. Which compound in the table is likely to be most mobile in a soil-groundwater system?

(A) 1,1-dichloroethene

(B) methylene chloride

(C) perchloroethene

(D) vinyl chloride

4.4. Which compound in the table is most likely to partition to the vapor phase?

(A) 1,1-dichloroethene

(B) methylene chloride

(C) perchloroethene

(D) vinyl chloride

4.5. Which compound in the table is most likely to exist as a nonaqueous phase liquid?

(A) 1,1-dichloroethene

(B) methylene chloride

(C) perchloroethene

(D) vinyl chloride

PROBLEM 5

5.1. What is the product of the photolytic decomposition of nitrogen dioxide in the atmosphere?

(A) nitric acid aerosol and particulate

(B) nitric oxide and free oxygen radicals

(C) photochemical smog

(D) stratospheric ozone scavenging peroxyl radicals

5.2. What is the cause of ground level ozone accumulation?

(A) photolytic decomposition of CO to produce free oxygen radicals

(B) photolytic oxidation of dissociated atmospheric water vapor

(C) peroxyl radical interruption of the normal NO_2 photolytic cycle

(D) peroxyl radical reaction with photolytically dissociated carbon dioxide

5.3. What compounds are predominantly involved in reactions with the hydroxyl radical (OH·) to form photochemical smog?

(A) carbon monoxide, aldehydes and ketones, and hydrocarbons

(B) carbon monoxide, nitrogen dioxide, and hydrocarbons

(C) carbon monoxide, nitrogen dioxide, and sulfur dioxide

(D) nitrogen dioxide, sulfur dioxide, and oxygen

5.4. What factor does not directly contribute to prevent the theoretically possible continual generation of ground level ozone?

(A) cloud shading and sun position

(B) wind and temperature gradients

(C) night/day cycle

(D) rainfall and cloud cover

5.5. Which reaction represents a terminating reaction to the sequence of reactions leading to the accumulation of ground level ozone?

(A) $O_3 + NO \longrightarrow NO_2$

(B) $HO_2\cdot + NO \longrightarrow NO_2 + OH\cdot$

(C) $OH\cdot + NO_2 \longrightarrow HNO_3$

(D) $H\cdot + O_2 \longrightarrow HO_2\cdot$

PROBLEM 6

A stream is contaminated with a compound that will bioaccumulate in fish tissue according to the single compartment model. The contaminant is present in the stream water at a concentration of 184 μg/L. The uptake rate constant is 1.32 mL/g·h, and the depuration rate constant is 1.43×10^{-3} h^{-1}. The fish are exposed over a period of 100 d.

6.1. What is the concentration of the contaminant in the fish tissue?

(A) 23 μg/g

(B) 150 μg/g

(C) 160 μg/g

(D) 200 μg/g

6.2. What is the bioconcentration factor?

(A) 0.87

(B) 29

(C) 870

(D) 1200

6.3. What is the contaminant half-life in the fish tissue?

(A) 0.24 d

(B) 4.0 d

(C) 20 d

(D) 100 d

6.4. How long is required for the contaminant to be completely eliminated from the fish tissue?

(A) 2.2 d

(B) 36 d

(C) 180 d

(D) 360 d

PROBLEM 7

Petroleum products released into the environment typically contaminate freshwater and marine systems. These released products are subject to a variety of dispersion and degradation mechanisms that are commonly referred to as weathering.

7.1. What is the primary mechanism responsible for weathering of heavier petroleum products, such as diesel fuel and lubricating oil, released to the environment?

(A) biodegradation

(B) dissolution

(C) photolysis

(D) volatilization

7.2. What is the primary mechanism responsible for weathering of lighter petroleum products, such as gasoline and aviation fuels, released to the environment?

(A) biodegradation

(B) dissolution

(C) photolysis

(D) volatilization

7.3. Which factors are most significant in reducing the degradation rate of petroleum products released to the marine environment?

(A) dissolution

(B) emulsification

(C) storm activity

(D) temperature increase

7.4. What distinguishes weathered petroleum products from those that are unweathered?

(A) Weathering decreases the amount of lower molecular weight compounds.

(B) Weathering decreases the bulk petroleum specific weight.

(C) Weathering decreases the viscosity of residual petroleum.

(D) Weathering increases the solubility of residual petroleum in water.

7.5. How does weathering affect toxicity of petroleum products released to the environment?

(A) Weathering decreases toxicity through the loss of more soluble compounds.

(B) Weathering decreases toxicity through increased mobility.

(C) Weathering increases toxicity through enrichment of more complex compounds.

(D) Weathering increases toxicity through the release of chloride and sulfur salts.

Remediation Solutions

1.1. $\mathrm{R_f D}_i$ = reference dose for each chemical i, mg/kg·d administered

NOAEL_i = no observed adverse effect level for chemical i, mg/kg·d

UF_1 = uncertainty factor for population effects
= 10

UF_2 = uncertainty factor for extrapolating animal data to humans
= 10

UF_3 = uncertainty factor for using NOAEL from subchronic instead of chronic studies
= 10

UF_4 = uncertainty factor for using LOAEL instead of NOAEL
= 1

SF = safety factor for other issues based on professional judgment
= 5

The administered oral reference dose is given by

$$\mathrm{R_f D}_i = \frac{\mathrm{NOAEL}_i}{(\mathrm{UF}_1)(\mathrm{UF}_2)(\mathrm{UF}_3)(\mathrm{UF}_4)(\mathrm{SF})}$$

$$\mathrm{R_f D}_1 = \frac{435\ \frac{\mathrm{mg}}{\mathrm{kg \cdot d}}}{(10)(10)(10)(1)(5)} = \boxed{0.087\ \mathrm{mg/kg \cdot d}}$$

$$\mathrm{R_f D}_2 = \frac{287\ \frac{\mathrm{mg}}{\mathrm{kg \cdot d}}}{(10)(10)(10)(1)(5)} = \boxed{0.057\ \mathrm{mg/kg \cdot d}}$$

$$\mathrm{R_f D}_3 = \frac{329\ \frac{\mathrm{mg}}{\mathrm{kg \cdot d}}}{(10)(10)(10)(1)(5)} = \boxed{0.066\ \mathrm{mg/kg \cdot d}}$$

The answer is C.

1.2. The oral reference dose absorbed via food is the product of the administered reference dose and the ingestion absorption efficiency.

The $\mathrm{R_f D}_1$ absorbed via food is

$$\left(\frac{0.087\ \mathrm{mg}}{\mathrm{kg \cdot d}}\right)\left(\frac{20\%\ \mathrm{absorbed}}{100\%\ \mathrm{administered}}\right)$$

$$= \boxed{0.017\ \mathrm{mg/kg \cdot d}}$$

The $\mathrm{R_f D}_2$ absorbed via food is

$$\left(\frac{0.057\ \mathrm{mg}}{\mathrm{kg \cdot d}}\right)\left(\frac{20\%\ \mathrm{absorbed}}{100\%\ \mathrm{administered}}\right)$$

$$= \boxed{0.011\ \mathrm{mg/kg \cdot d}}$$

The $\mathrm{R_f D}_3$ absorbed via food is

$$\left(\frac{0.066\ \mathrm{mg}}{\mathrm{kg \cdot d}}\right)\left(\frac{20\%\ \mathrm{absorbed}}{100\%\ \mathrm{administered}}\right)$$

$$= \boxed{0.013\ \mathrm{mg/kg \cdot d}}$$

The answer is B.

1.3. Assume the dose absorbed via food and water is the same. The administered oral reference dose equals the absorbed reference dose divided by the ingestion absorption efficiency.

The $\mathrm{R_f D}_1$ administered via water is

$$\left(\frac{0.017\ \mathrm{mg}}{\mathrm{kg \cdot d}}\right)\left(\frac{100\%\ \mathrm{administered}}{90\%\ \mathrm{absorbed}}\right)$$

$$= \boxed{0.019\ \mathrm{mg/kg \cdot d}}$$

The $\mathrm{R_f D}_2$ administered via water is

$$\left(\frac{0.011\ \mathrm{mg}}{\mathrm{kg \cdot d}}\right)\left(\frac{100\%\ \mathrm{administered}}{90\%\ \mathrm{absorbed}}\right)$$

$$= \boxed{0.012\ \mathrm{mg/kg \cdot d}}$$

The $\mathrm{R_f D}_3$ administered via water is

$$\left(\frac{0.013\ \mathrm{mg}}{\mathrm{kg \cdot d}}\right)\left(\frac{100\%\ \mathrm{administered}}{90\%\ \mathrm{absorbed}}\right)$$

$$= \boxed{0.014\ \mathrm{mg/kg \cdot d}}$$

The answer is B.

1.4. HI_i = hazard index for each chemical i
  I_i = intake for each chemical i, mg/kg·d
  R_fD_i = absorbed oral reference dose for each chemical i through ingestion of water, mg/kg·d

$$HI_i = \frac{I_i}{R_fD_i}$$

$$HI_1 = \frac{0.020 \ \frac{mg}{kg \cdot d}}{0.017 \ \frac{mg}{kg \cdot d}} = \boxed{1.18 \quad (1.2)}$$

$$HI_2 = \frac{0.0092 \ \frac{mg}{kg \cdot d}}{0.011 \ \frac{mg}{kg \cdot d}} = \boxed{0.84}$$

$$HI_3 = \frac{0.014 \ \frac{mg}{kg \cdot d}}{0.013 \ \frac{mg}{kg \cdot d}} = \boxed{1.08 \quad (1.1)}$$

The hazard index for chemicals 1 and 3 is greater than 1.0; therefore, these two chemicals should be targeted for remediation.

The answer is C.

SOLUTION 2

2.1. For herbicide 1, the 96 h LC_{50} is

$$1800 \ \frac{\mu g}{L} - \frac{\left(1800 \ \frac{\mu g}{L} - 1600 \ \frac{\mu g}{L}\right)(50\% - 46\%)}{58\% - 46\%}$$

$$= \boxed{1733 \ \mu g/L}$$

For herbicide 2, the 96 h LC_{50} is

$$280 \ \frac{\mu g}{L} - \frac{\left(280 \ \frac{\mu g}{L} - 260 \ \frac{\mu g}{L}\right)(50\% - 44\%)}{53\% - 44\%}$$

$$= \boxed{267 \ \mu g/L}$$

The answer is C.

2.2. Herbicide 2 is more toxic than herbicide 1 because herbicide 2 has a lower 96 h LC_{50} than does herbicide 1.

The answer is C.

2.3. For herbicide 1, the bioconcentration factor is

$$\frac{38 \ \frac{\mu g}{kg}}{\left(0.9 \ \frac{\mu g}{L}\right)\left(1 \ \frac{L}{kg}\right)} = \boxed{42}$$

For herbicide 2, the bioconcentration factor is

$$\frac{64 \ \frac{\mu g}{kg}}{\left(0.1 \ \frac{\mu g}{L}\right)\left(1 \ \frac{L}{kg}\right)} = \boxed{640}$$

The answer is C.

SOLUTION 3

3.1. The arsenic is likely to be present in the drinking water as the more toxic arsenite (As(III)) because the groundwater dissolved oxygen concentration of 0.8 mg/L suggests reducing conditions.

The answer is D.

3.2. C = concentration = 109 $\mu g/L$
  ED = exposed duration
    = 70 yr (for lifetime resident)
  %T = % of time exposed as fraction (assume 1)
  SF = slope factor = $5 \times 10^{-5} \ (\mu g/L)^{-1}$
  LT = lifetime
    = 70 yr (standard USEPA exposure value)

$$risk = \frac{C(ED)(\%T)(SF)}{LT}$$

$$= \frac{\left(109 \ \frac{\mu g}{L}\right)((70 \ yr)(1))\left(5 \times 10^{-5} \ \left(\frac{\mu g}{L}\right)^{-1}\right)}{70 \ yr}$$

$$= \boxed{0.0055 \ or \ 55 \ in \ 10^4}$$

Acceptable cancer risk is typically considered to be 1 in 10^6 or less, therefore, 55 in 10^4 would be considered unacceptable. The residents are likely to be exposed to a significant cancer risk from ingesting the arsenic in their drinking water.

The answer is D.

3.3. HI = hazard index, unitless

DI = daily intake

$\quad$ = 2 L/d (standard USEPA exposure value)

%A = percent absorbed (assume 1)

%T = percent time exposed (assume 1)

BW = body weight

$\quad$ = 70 kg (standard USEPA exposure value)

R_fD = reference dose = 1×10^{-3} mg/kg·d

$$HI = \frac{C(DI)(\%A)(ED)(\%T)}{(BW)(LT)(R_fD)}$$

$$= \frac{\left(109 \, \frac{\mu g}{L}\right)\left(\frac{1 \text{ mg}}{1000 \, \mu g}\right)\left(\left(2 \, \frac{L}{d}\right)(1)\right)((70 \text{ yr})(1))}{(70 \text{ kg})(70 \text{ yr})\left(1 \times 10^{-3} \, \frac{\text{mg}}{\text{kg·d}}\right)}$$

$$= 3.1$$

When HI is greater than 1.0, potential noncarcinogenic health hazard concerns exist. Since HI of 3.1 exceeds 1.0, the residents are likely to be exposed to a significant noncarcinogenic health risk from ingesting the arsenic in their drinking water.

The answer is D.

3.4. The bioconcentration factor is developed for exposures to fish, not direct ingestion of contaminated drinking water by humans. However, it does suggest that arsenic is absorbed by biological tissue. Considering the exposed dose through ingestion of drinking water and other exposure through ingestion of vegetables irrigated with the contaminated water, it is likely that some bioaccumulation of arsenic will occur in the residents' bodies over a lifetime of exposure.

The answer is C.

3.5. R_fD = reference dose = 6×10^{-2} mg/kg·d

$$HI = \frac{\left(1.7 \, \frac{\text{mg}}{L}\right)\left(\left(2 \, \frac{L}{d}\right)(1)\right)((70 \text{ yr})(1))}{(70 \text{ kg})(70 \text{ yr})\left(6 \times 10^{-2} \, \frac{\text{mg}}{\text{kg·d}}\right)} = 0.8$$

The HI of 0.8 is less than 1.0, indicating that no likely health hazard exists. Also, given that municipal water supplies are commonly fluoridated to provide a fluoride concentration of 1.0 mg/L and that the MCL for fluoride is 4 mg/L, a concentration of 1.7 mg/L will not likely present any significant negative health consequences to the residents from ingesting the fluoride with the drinking water.

The answer is A.

SOLUTION 4

4.1. Biological oxidation, chemical reduction, and hydrolysis are possible significant pathways for degradation of perchloroethene in a soil-groundwater system, but deamination is not.

The answer is C.

4.2. The degradation sequence from the dichloroethene isomers would be to a monochloroethene. Therefore, each dichloroethene will degrade to vinyl chloride (CH_2-$CHCl$).

The answer is D.

4.3. Mobility in a soil-groundwater system generally increases with decreasing values of the soil-water partition coefficient. The compound in the table with the lowest soil-water partition coefficient is methylene chloride.

The answer is B.

4.4. The potential to partition to the vapor phase, or volatility, increases with increasing values of Henry's constant. The compound in the table with the greatest Henry's constant is vinyl chloride.

The answer is D.

4.5. As solubility decreases, the potential for a compound to exist as a nonaqueous phase liquid increases. The compound in the table with the lowest solubility is perchloroethene.

The answer is C.

SOLUTION 5

5.1. The normal NO_2 photolytic cycle is represented by the following reaction sequence.

$$NO_2 + h\nu \longrightarrow NO + O\cdot$$
$$O\cdot + O_2 + N_2 \longrightarrow O_3 + N_2$$
$$O_3 + NO \longrightarrow NO_2 + O_2$$

The reaction shows that the photolytic decomposition of nitrogen dioxide in the atmosphere produces nitric oxide and free oxygen radicals. Subsequent reactions produce ozone, which then reacts with nitric oxide to consume the ozone and produce nitrogen dioxide and O_2, completing the cycle.

The answer is B.

5.2. When the normal NO_2 photolytic cycle is interrupted by peroxyl radicals, the reaction of O_3 with NO is terminated and ground level ozone accumulates.

The answer is C.

5.3. Carbon monoxide, aldehydes and ketones (carbonyls), and hydrocarbons are predominantly involved in reactions with the hydroxyl radical (OH·) to form photochemical smog. If these three groups of compounds are absent, photochemical smog will not be present.

The answer is A.

5.4. Cloud shading and sun position, night/day cycle, and rainfall and cloud cover all directly contribute to prevent theoretically possible continual generation of ground level ozone. However, wind and temperature gradients do not.

The answer is B.

5.5. $O_3 + NO \rightarrow NO_2$ is the final reaction in the normal NO_2 photolytic cycle and is not part of the chain of reactions leading to ground level ozone accumulation. $HO_2· + NO \rightarrow NO_2 + OH·$ and $H· + O_2 \rightarrow HO_2·$ each produce a radical that is involved in subsequent reactions that contribute to ground level ozone accumulation. $OH· + NO_2 \rightarrow HNO_3$ is a terminating reaction where the hydroxyl radical is removed from play by the formation of nitric acid.

The answer is C.

SOLUTION 6

6.1. C_b = contaminant concentration in biota (fish), $\mu g/g$
C_m = contaminant concentration in medium (water)
$\quad = 184\ \mu g/L$
K_u = rate coefficient for contaminant uptake
$\quad = 1.32\ mL/g\cdot h$
K_d = rate coefficient for contaminant depuration
$\quad = 1.43 \times 10^{-3}\ h^{-1}$

$$t = (100\ d)\left(24\ \frac{h}{d}\right) = 2400\ h$$

$$C_b = C_m\left(\frac{K_u}{K_d}\right)(1 - e^{-tK_d})$$

$$= \frac{\left(\begin{array}{l}\left(184\ \frac{\mu g}{L}\right)\left(\dfrac{1.32\ \frac{mL}{g\cdot h}}{1.43 \times 10^{-3}\ h^{-1}}\right) \\[6pt] \times \left(1 - e^{-(2400\ h)(1.43\times 10^{-3}\ h^{-1})}\right)\end{array}\right)}{1000\ \dfrac{mL}{L}}$$

$$= \boxed{164\ \mu g/g \quad (160\ \mu g/g)}$$

The answer is C.

6.2. BCF = bioconcentration factor, unitless
ρ_m = medium density = 1000 g/L

$$\text{BCF} = \frac{C_b}{C_m(\rho_m)^{-1}}$$

$$= \frac{\left(160\ \frac{\mu g}{g}\right)\left(1000\ \frac{g}{L}\right)}{184\ \frac{\mu g}{L}}$$

$$= \boxed{870}$$

The answer is C.

6.3. $t_{1/2}$ = contaminant half-life in the fish tissue

$$t_{1/2} = \frac{0.693}{K_d}$$

$$= \frac{(0.693)\left(\frac{1\ d}{24\ h}\right)}{1.43 \times 10^{-3}\ h^{-1}}$$

$$= \boxed{20.2\ d \quad (20\ d)}$$

The answer is C.

6.4. The usual assumption is that the contaminant will be completely eliminated from the fish tissue following nine half-lives.

The time required for the contaminant to be eliminated from the fish tissue is

$$9t_{1/2} = (9)(20\ d) = \boxed{180\ d}$$

The answer is C.

SOLUTION 7

7.1. Biodegradation is the primary mechanism responsible for weathering of heavier petroleum products, such as diesel fuel and lubricating oil, released to the environment. Photolysis may also occur if heavy petroleum products are exposed to direct sunlight (but to a lesser degree), and dissolution and volatilization are minor mechanisms, if they occur at all.

The answer is A.

7.2. The primary mechanism responsible for weathering of lighter petroleum products, such as gasoline and aviation fuels, released to the environment is volatilization to the atmosphere or, if the release is subsurface, to soil vapor. Dissolution may also occur, but would not be as significant as volatilization. Volatilization of lighter fractions would occur relatively rapidly and allow less time for significant biodegradation or, if exposed to direct sunlight, photolysis.

The answer is D.

7.3. Degradation of petroleum products released to the marine environment is most significant through biological, photolytic, and vaporization pathways, all of which are enhanced by surface spreading of the hydrocarbon to increase surface area. Consequently, factors such as emulsification that reduce surface area are most significant in reducing the degradation rate of petroleum products released to the marine environment.

The answer is B.

7.4. Weathered petroleum products are most readily distinguished from those that are unweathered by the loss of lower molecular weight compounds that are more easily volatilized, biodegraded, and dissolved.

The answer is A.

7.5. Weathering reduces the toxicity of petroleum products released to the environment by the loss of the more water soluble, lower molecular weight compounds.

The answer is A.

Public Health and Safety

The OSHA Hazardous Waste Operations and Emergency Response (HAZWOPER) Standard in 29 CFR 1910.120 is intended to limit employee exposure to safety and health hazards at TSD facilities during clean-up and corrective actions at contaminated sites and during other emergency response activities involving hazardous wastes and substances.

1.1. What constitutes an emergency response?

(A) a response effort by persons outside the immediate release area to an uncontrolled release of a hazardous substance

(B) a response effort by persons at the immediate release area where the hazardous substances can be controlled at the time of release

(C) a response effort by persons at the immediate release area where the release does not present a potential safety or health hazard

(D) all of the above

1.2. What does IDLH describe?

(A) that concentration of oxygen by volume below which atmosphere-supplying respiratory protection must be provided

(B) an atmospheric concentration of any toxic, corrosive, or asphyxiant substance that poses an immediate threat to life or would interfere with a person's ability to escape from a dangerous atmosphere

(C) the permissible exposure limit through inhalation or dermal contact as published in the regulation or by NIOSH or CGIH

(D) all of the above

1.3. Who are first responders at the awareness level?

(A) the first-arriving most senior emergency response official in charge of a site through which all emergency responders and their communications are coordinated

(B) the person knowledgeable in the operations being implemented at an emergency response site who has specific responsibility to identify and evaluate hazards and to provide supervision

(C) employees who in the course of their regular job duties work with and are trained in the hazards of specific substances and are called to advise other emergency responders

(D) individuals who are likely to witness or discover a hazardous substance release and who have been trained to initiate an emergency response sequence by notifying the authorities of the release

1.4. How many hours of off-site training are required for general site workers?

(A) 8

(B) 16

(C) 24

(D) 40

1.5. What actions are process operators authorized to perform in response to a hazardous substance release?

(A) no action beyond notifying the emergency response team of the release

(B) no action unless the emergency response team is prevented from accessing the release area

(C) only the limited action specifically addressed in the emergency response plan

(D) any first-responder defensive action unless specifically prevented by the emergency response plan

The following questions pertain to the OSHA Hazard Communication Standard defined under 29 CFR 1910.1200.

2.1. Which of the following is not part of the Hazard Communication Standard?

(A) container labeling

(B) personal exposure monitoring

(C) material safety data sheets

(D) employee training

2.2. How is a combustible liquid defined under the Hazard Communication Standard?

(A) flash point below 38°C (100°F)

(B) flash point between 38°C (100°F) and 93°C (200°F)

(C) flash point below 60°C (140°F)

(D) flash point below 100°C (212°F)

2.3. Who is responsible for evaluating chemicals to determine if they are hazardous?

(A) employers only

(B) employers and manufacturers

(C) manufacturers only

(D) manufacturers and importers

2.4. When are employers required to provide training to employees on hazardous chemicals in their work areas?

(A) when employees are initially assigned to a work area

(B) within 15 d of the introduction of a new health hazard into the employees' work area

(C) within 15 d of an incident involving an unacceptable exposure of employees to hazardous chemicals

(D) quarterly, but only when material safety data sheets are not available for hazardous chemicals used by employees

2.5. Which is not a basis for designating a chemical as a health hazard under the Hazard Communication Standard?

(A) The chemical is corrosive.

(B) The chemical is an irritant.

(C) The chemical is reactive.

(D) The chemical is a sensitizer.

PROBLEM 3

A workplace exposure survey has produced the following results.

	acetone	toluene
1 h exposure at concentration (ppm)	1210	NA
2 h exposure at concentration (ppm)	719	NA
3 h exposure at concentration (ppm)	NA	380
5 h exposure at concentration (ppm)	370	140
8 h TWA PEL* (ppm)	1000	200
acceptable ceiling (ppm)	NA	300
acceptable maximum peak above ceiling		
concentration (ppm)	NA	500
duration (min)	NA	10

*TWA PEL is the time-weighted average peak exposure limit.

3.1. What is the cumulative exposure to acetone?

(A) 560 ppm

(B) 770 ppm

(C) 1640 ppm

(D) 2300 ppm

3.2. What is the cumulative exposure to the acetone and toluene mixture?

(A) 0.48

(B) 1.7

(C) 2.4

(D) 4.9

3.3. Is the cumulative exposure to the acetone and toluene mixture acceptable?

(A) Yes, because the cumulative exposure for the mixture is less than 1.0.

(B) Yes, because the cumulative exposure for the mixture is greater than 1.0.

(C) No, because the cumulative exposure for the mixture is less than 1.0.

(D) No, because the cumulative exposure for the mixture is greater than 1.0.

3.4. If a peak toluene exposure of 480 ppm occurs for 10 min during the first 3 h, what is the cumulative exposure to toluene?

(A) 230 ppm

(B) 330 ppm

(C) 1000 ppm

(D) 3000 ppm

3.5. If a peak toluene exposure of 480 ppm occurs for 10 min during the first 3 h, is the cumulative exposure to toluene acceptable?

(A) Yes, because the cumulative exposure for toluene is greater than 1.0.

(B) No, because the cumulative exposure for toluene is less than 1.0.

(C) Yes, because the cumulative exposure for toluene is less than the TWA PEL for toluene.

(D) No, because the cumulative exposure for toluene is greater than the TWA PEL for toluene.

PROBLEM 4

The Uniform Fire Code (UFC) provides guidance for classifying and handling a wide variety of materials. The following questions are from the UFC.

4.1. Which of the following is not included in general requirements for hazardous materials storage?

(A) compatibility among stored materials

(B) setbacks from mechanical equipment

(C) spill control and containment

(D) storage cabinet construction

4.2. For an aboveground tank with a capacity of less than 12,000 gal that is used to store a Class III-B combustible liquid, what is the minimum setback from the property line of a lot with a building?

(A) 5 ft

(B) 10 ft

(C) 15 ft

(D) 20 ft

4.3. What is the description of a Class IV organic peroxide?

(A) does not burn or present a decomposition hazard

(B) burns as ordinary combustible and presents a minimum reactivity hazard

(C) burns rapidly and presents a moderate reactivity hazard

(D) burns very rapidly and presents a severe reactivity hazard

4.4. What is the description of a Class I unstable, non-water reactive material?

(A) readily capable of detonation

(B) capable of detonation

(C) normally unstable, but does not detonate

(D) normally stable

4.5. Which of the following is not a classification for cryogenic fluids?

(A) corrosive/highly toxic

(B) flammable

(C) nonflammable

(D) reactive/unstable

PROBLEM 5

The following questions address OSHA Standard 1910.119, Process Safety Management (PSM) of Highly Hazardous Chemicals.

5.1. The PSM standard applies to a flammable liquid or gas in a quantity of 10,000 lbm or more except under what condition?

(A) Any employee using the material does so only while wearing the appropriate personal protective equipment.

(B) The material is stored in an underground tank or in containers inside a dedicated aboveground structure.

(C) The material is used exclusively for workplace consumption as a fuel.

(D) The material is kept below its normal boiling point by employing chilling or refrigeration.

5.2. What type of facilities are not subject to the PSM standard?

(A) normally unoccupied remote facilities

(B) oil or gas well drilling or servicing operations

(C) retail facilities

(D) all of the above

5.3. At what interval after completing the initial process hazard analysis does the process hazard analysis need to be updated and revalidated?

(A) 3 yr

(B) 5 yr

(C) 2 yr after the initial analysis and 3 yr thereafter

(D) 2 yr after the initial analysis and 5 yr thereafter

5.4. How frequently should employee refresher training be provided under the PSM standard?

(A) as needed

(B) as needed, but at least every 1 yr

(C) as needed, but at least every 2 yr

(D) as needed, but at least every 3 yr

5.5. What conditions prompt a compliance audit under the PSM standard?

(A) A regulated material is released in the workplace.

(B) An employee is injured in an area where regulated materials are used.

(C) Three years have elapsed since the previous audit.

(D) Turnover of trained employees exceeds 20%.

PROBLEM 6

Requirements for the selection and use of personal protective equipment (PPE) in the workplace are stipulated in 29 CFR 1910 Subpart I with special requirements under CERCLA identified in 40 CFR 300.

6.1. What responsibilities do employers have when employees provide their own PPE?

(A) The employer has no responsibility for employee equipment.

(B) The employer must ensure that the equipment is adequate.

(C) The employer must ensure that the equipment is adequate, including proper maintenance.

(D) The employer must ensure that the equipment is adequate, including proper maintenance and sanitation.

6.2. What responsibility does the employer have for ensuring that employees are proficient in the use of PPE?

(A) The employer must provide training.

(B) The employer must provide training and require the employee to demonstrate proficiency.

(C) The employer must provide training, require the employee to demonstrate proficiency, and retrain employees who lose proficiency.

(D) The employer must provide training, require the employee to demonstrate proficieny, retrain employees who lose proficiency, and provide written certification.

6.3. What type of respirator would be suitable for entering a space that is immediately dangerous to life and health (IDLH)?

(A) positive-pressure air-line respirator

(B) positive-pressure air-purifying respirator

(C) positive-pressure self-contained breathing apparatus

(D) positive-pressure supplied-air respirator

6.4. Where chemical-protective clothing (CPC) is required, under what conditions would CPC with a high "clo" value be most suitable?

(A) lower work rate and lower temperature

(B) lower work rate and higher temperature

(C) higher work rate and lower temperature

(D) higher work rate and higher temperature

6.5. What USEPA level of protection defines the highest degree of respiratory, skin, and eye protection?

(A) level A

(B) level B

(C) level C

(D) level D

6.6. Which of the following is probably not a factor in evaluating a worker's susceptibility to heat stress when wearing chemical-protective clothing?

(A) age

(B) gender

(C) health

(D) physical fitness

PROBLEM 7

A radiation exposure through ingestion results in an alpha particle absorbed dose of 0.012 rad and a beta particle absorbed dose of 0.32 rad. The quality factors for alpha and beta particles are 20 and 1, respectively.

7.1. What is the dose-equivalent from exposure to the alpha radiation?

(A) 0.012 rad

(B) 0.24 rem

(C) 0.32 rad

(D) 0.32 rem

7.2. What is the dose-equivalent from exposure to the beta radiation?

(A) 0.012 rad

(B) 0.24 rem

(C) 0.32 rad

(D) 0.32 rem

7.3. Does exposure to the alpha radiation or to the beta radiation present a greater hazard to the exposed individual?

(A) exposure to alpha radiation, because alpha particles have more energy than beta particles

(B) exposure to beta radiation, because beta particles have more energy than alpha particles

(C) exposure to alpha radiation, because it occurs at a greater dose-equivalent

(D) exposure to beta radiation, because it occurs at a greater dose-equivalent

7.4. What alpha particle absorbed dose would be equivalent to the given beta particle absorbed dose?

   (A) 0.012 rad

   (B) 0.016 rad

   (C) 0.24 rem

   (D) 0.32 rem

PROBLEM 8

The following questions relate to radiation and its health effects.

8.1. What units are used to express radioactivity?

   (A) rem

   (B) rad

   (C) curie

   (D) roentgen

8.2. In what order would travel distances of radiation through air increase?

   (A) alpha < beta < gamma

   (B) beta < alpha < gamma

   (C) gamma < beta < alpha

   (D) gamma < alpha < beta

8.3. How are "absorbed dose" and "dose-equivalent" related?

   (A) Absorbed dose and dose-equivalent are synonymous terms used to describe radiation energy absorbed per unit time.

   (B) Dose-equivalent is the product of absorbed dose and a unitless conversion factor specific to the type of radioactivity.

   (C) Dose-equivalent is the sum of absorbed doses for all types of radioactivity to which exposure occurs.

   (D) They are unrelated; dose-equivalent defines relative particle radioactivity and absorbed dose represents exposure.

8.4. Through which exposure pathway(s) do alpha particles present the greatest health hazard?

   (A) inhalation

   (B) inhalation and ingestion

   (C) ingestion and dermal

   (D) dermal

8.5. What type of protective clothing will shield individuals from gamma radiation?

   (A) cotton long-sleeve shirt, long pants, cap, gloves

   (B) heavy, multi-layered canvas suit with gloves, hood, and plastic face shield

   (C) lead-sheet lined, full-body suit with leaded-glass eye shield

   (D) no protective clothing that will still allow the individual to move about

8.6. For unprotected individuals, do radiation contamination and radiation exposure always occur together?

   (A) contamination and exposure are the same thing

   (B) although contamination and exposure are different, one cannot occur without the other

   (C) radiation exposure can occur without contamination

   (D) radiation contamination can occur without exposure

PROBLEM 9

The following problems address electromagnetic radiation composed of nonionizing and ionizing radiation.

9.1. Which of the following is not an example of non-ionizing radiation?

   (A) medical diagnostic x-ray

   (B) microwave

   (C) radio wave

   (D) ultraviolet light

9.2. What is the greatest worldwide contributor to ionizing radiation exposure?

   (A) defense industry and stockpiles

   (B) medical applications including x-rays

   (C) natural background sources

   (D) nuclear reactor wastes (all sources)

9.3. Which factor exerts the most overall influence on cosmic radiation exposure at the earth's surface?

   (A) altitude

   (B) dress and outdoor activity

   (C) latitude

   (D) weather

9.4. What are the three major series of radioactive elements?

(A) actinium, polonium, thorium

(B) actinium, thorium, uranium

(C) polonium, radium, thorium

(D) radium, thorium, uranium

9.5. What is the effective half-life of a nuclide with a physical half-life of 15 h and a biological half-life of 11 d?

(A) 15 h

(B) 249 h

(C) 11 d

(D) 17.6 h/h

PROBLEM 10

The following problems address hazardous chemical reporting pertaining to community right-to-know issues defined under 40 CFR 370.

10.1. Which of the following is not included as a hazard category?

(A) an immediate (acute) health hazard

(B) a delayed (chronic) health hazard

(C) corrosives, other than immediate health hazards

(D) a sudden release of pressure

10.2. What is the basic requirement for MSDS reporting to the local emergency planning committee?

(A) Regulated facilities shall submit an MSDS for each hazardous chemical present at a facility.

(B) Regulated facilities shall submit an MSDS only for each hazardous material stored at an outside location at a facility.

(C) Regulated facilities shall submit an MSDS only for each hazardous material that is a precursor to a hazardous waste stream.

(D) Regulated facilities shall submit an MSDS only for materials that present a health hazard when released, discharged, or emitted.

10.3. What information is required on Tier II chemical inventory forms that is not required on Tier I forms?

(A) Tier II forms include information describing physical and health hazards of chemicals included in the inventory and Tier I forms do not.

(B) Tier II forms require certification by the facility owner/operator or his representative and Tier I forms do not.

(C) Tier II forms include specific information listed by chemical such as chemical names and CAS numbers and Tier I forms do not.

(D) Tier II forms require an emergency contact name and 24 h telephone number and Tier I forms do not.

10.4. Who may request MSDS information for a regulated facility and expect a response from the local emergency planning committee?

(A) any person

(B) only persons designated by the local emergency planning committee

(C) only potential emergency responders to the facility

(D) only persons designated by the local emergency planning committee and potential emergency responders to the facility

10.5. What penalties may result from failure of a regulated facility to comply with MSDS or inventory reporting requirements?

(A) The regulation does not specify penalties.

(B) unspecified administrative penalties

(C) monetary civil and administrative penalties

(D) unspecified civil and criminal penalties

PROBLEM 11

Emergency planning and notification are required for facilities at which an amount of an extremely hazardous substance or CERCLA hazardous substance is present in excess of the regulated quantity. Emergency planning and notification requirements are specified in 40 CFR 355.

11.1. How are "CERCLA hazardous substance," "extremely hazardous substance," and "hazardous chemical" differentiated?

(A) The terms are synonymous.

(B) Extremely hazardous substances and hazardous chemicals make up subgroupings of CERCLA hazardous substances.

(C) Each term defines specific groups of materials under different regulatory citations.

(D) CERCLA hazardous substances and extremely hazardous substances are synonyms for materials regulated under CERCLA, and hazardous chemicals are regulated under OSHA.

11.2. What is "threshold planning quantity" (TPQ)?

(A) Threshold planning quantity is equal to reportable quantity.

(B) the quantity of a material handled at a facility that, when equaled or exceeded, subjects the facility to the regulation

(C) the quantity of regulated material that, when released, requires notification of emergency response agencies

(D) the quantity of a regulated material that a facility must be prepared to contain if a release occurs

11.3. What do the two values listed for threshold planning quantity designate?

(A) The first value applies to facilities with less than 100 employees, and the second value applies to facilities with greater than 100 employees.

(B) The first value applies to facilities located within 1000 ft of sensitive population areas, and the second value applies to facilities located away from these areas.

(C) The first value applies to materials with higher hazard properties, and the second value applies to materials with lower hazard properties.

(D) The first value applies to gases and liquids only, and the second value applies to solids only.

11.4. When is emergency response notification not required?

(A) when the release occurs as abandonment of intact closed containers such as barrels

(B) when the release results in exposure to persons solely within the boundaries of the facility

(C) when the release does not involve a CERCLA hazardous substance

(D) when the release is less than the threshold planning quantity

11.5. What information is not required to be provided as part of emergency release notification?

(A) the identity or name of the released material

(B) an estimate of the quantity of the released material

(C) the name and title of the person reporting the release

(D) the time and duration of the release

11.6. What penalties may result from failure to report a release subject to the regulations?

(A) unspecified administrative penalties

(B) unspecified civil and criminal penalties

(C) monetary civil and administrative penalties

(D) monetary civil and criminal penalties or imprisonment

PROBLEM 12

The National Fire Protection Association (NFPA) and others have defined standards and guidelines for fire safety. The following problems address issues related to these standards and guidelines.

12.1. What types of materials are involved in an NFPA Class B fire?

(A) wood, paper, cloth

(B) oil, gasoline, paint

(C) live electrical equipment

(D) combustible metals

12.2. What flash point defines an NFPA Class II combustible liquid?

(A) greater than or equal to 73°F (23°C) but less than 100°F (38°C)

(B) greater than or equal to 100°F (38°C) but less than 140°F (60°C)

(C) greater than or equal to 140°F (60°C) but less than 200°F (93°C)

(D) greater than or equal to 200°F (93°C)

12.3. What term generally describes a rapid pressure increase in a confined vessel that results in a sudden release as the vessel ruptures?

(A) combustion

(B) deflagration

(C) detonation

(D) explosion

12.4. Foam agents are suitable for suppressing what NFPA fire classes?

(A) Class A only

(B) Class A and B

(C) Class B and C

(D) Class D only

12.5. A fire involving materials from which of the following chemical categories would likely not be appropriately suppressed using water?

(A) corrosive

(B) oxidizing

(C) reactive

(D) unstable

PROBLEM 13

The following problems address issues associated with noise and noise pollution.

13.1. What are the most important parameters commonly used to characterize sound?

(A) pressure, duration, intensity

(B) pressure, power, intensity

(C) power, duration, frequency

(D) intensity, duration, frequency

13.2. What is the common unit of noise?

(A) decibel (dB)

(B) sone

(C) phon

(D) hertz (Hz)

13.3. What is the audible frequency range for humans?

(A) between 16 Hz and 1000 Hz

(B) between 16 Hz and 10 000 Hz

(C) between 16 Hz and 20 000 Hz

(D) between 16 Hz and 30 000 Hz

13.4. Which of the following is typically not considered a direct noise attenuating factor?

(A) precipitation and fog

(B) wind and temperature gradients

(C) solar radiation

(D) molecular absorption

13.5. What noise source is generally associated with perceived noise level (PNL), composite noise rating (CNR), and noise exposure forecast (NEF)?

(A) airport noise

(B) community noise

(C) industrial noise

(D) urban traffic noise

PROBLEM 14

Noise at a construction site located approximately 100 m from a residential area is characterized by a sound power of 50 W and a frequency of 1000 Hz.

14.1. What is the sound power level?

(A) 100 dB

(B) 120 dB

(C) 130 dB

(D) 140 dB

14.2. What is the sound intensity in the residential area?

(A) 4.0×10^{-4} W/m^2

(B) 1.6×10^{-3} W/m^2

(C) 5.0×10^{-3} W/m^2

(D) 6.4×10^{-3} W/m^2

14.3. What is the sound intensity level in the residential area?

(A) 86 dB

(B) 92 dB

(C) 96 dB

(D) 98 dB

14.4. What is the sound pressure level in the residential area?

(A) 86 dB

(B) 92 dB

(C) 96 dB

(D) 98 dB

14.5. What is the loudness in the residential area?

(A) 24 sones

(B) 37 sones

(C) 56 sones

(D) 64 sones

PROBLEM 15

The following problems apply to ISO 14000.

15.1. What is ISO 14000?

(A) mandatory environmental standards imposed upon member countries of the World Trade Organization

(B) mandatory environmental standards imposed upon private-sector business and industry importing to member countries of the European Union

(C) voluntary environmental standards applied to private-sector European and North American trading partners

(D) voluntary environmental standards applied to private-sector business and industry worldwide

15.2. What are the two general categories into which ISO 14000 standards are divided?

(A) organization evaluation standards and product and process evaluation standards

(B) organization evaluation standards and media release/discharge/emission standards

(C) product and process evaluation standards and waste classification and characteristics standards

(D) waste classification and characteristics standards and media release/discharge/emission standards

15.3. Which is not included as a major part of the ISO 14000 standards?

(A) environmental auditing (EA)

(B) environmental management systems (EMS)

(C) environmental site assessment (ESA)

(D) life cycle assessment (LCA)

15.4. What are the three possible outcomes of the ISO 14000 registration process?

(A) compliance, enforcement, penalty

(B) approval, conditional approval, disapproval

(C) certified, compliant, noncompliant

(D) permitted, registered, listed

PROBLEM 16

The following problems apply to the Toxic Substances and Control Act (TSCA) and the Federal Insecticide, Fungicide, and Rodenticide Act (FIFRA).

16.1. What chemicals must appear on the TSCA inventory list?

(A) all chemicals assigned a CAS number

(B) all chemicals manufactured for sale

(C) only chemicals exhibiting the hazardous waste characteristics of toxicity, corrosivity, ignitability, or reactivity

(D) only chemicals with a USEPA hazardous waste number

16.2. What requirements does FIFRA place on registration of pesticides?

(A) Only pesticides registered with the USEPA can be distributed, sold, or placed into commerce.

(B) Only pesticides registered with the U.S. Department of Agriculture can be distributed, sold, or placed into commerce.

(C) Only pesticides exhibiting characteristics of "persistence" and "lipid solubility" as defined by the Act need be registered.

(D) Only pesticides sold for commercial agriculture and commercial pest control applications need be registered.

16.3. What defines general-use pesticides?

(A) pesticides available to certified private applicators for general agriculture uses, but only related to the applicator's specific agricultural operation

(B) pesticides available for distribution and sale only to certified commercial applicators for general pest control activities

(C) pesticides that are not used on food chain crops, or that do not leave a residue on food chain crops after harvesting, and that do not present unreasonable adverse health effects to farm workers

(D) pesticides that can be used by untrained persons according to label instructions without creating unreasonable adverse effects to the person applying the pesticide or to the environment

PROBLEM 17

Life safety is addressed by NFPA 101 in the Life Safety Code® and by the Uniform Building Code (UBC). The following problems address life safety issues.

17.1. Which of the following is covered by the Life Safety Code®?

(A) building services and fire protection equipment

(B) personal protective equipment

(C) process safety management

(D) hazardous materials compatibility and segregation

17.2. What hazard of contents classification is employed in the Life Safety Code®?

(A)   corrosive, flammable, oxidizer, and reactive

(B)   low, ordinary, and high

(C)   nonflammable, flammable, combustible, and explosive

(D)   unclassified, Class I, and Class II

17.3. What topics are addressed under means of egress in either the Life Safety Code® or the UBC?

(A)   interior finishes and floor coverings

(B)   staircase dimensional criteria

(C)   smoke control and smoke and heat venting

(D)   features of fire protection equipment

17.4. What is the UBC group classification for hazardous occupancy?

(A)   Group B

(B)   Group F

(C)   Group H

(D)   Group M

17.5. What is the UBC classification for interior finishes with a flame-spread index of less than 25?

(A)   unclassified

(B)   Class I

(C)   Class II

(D)   Class III

Public Health and Safety Solutions

SOLUTION 1

1.1. An emergency response is defined in 29 CFR 1910.120(a)(3) as a response effort by persons outside the immediate release area to an uncontrolled release of a hazardous substance.

The answer is A.

1.2. In 29 CFR 1910.120(a)(3), IDLH (immediately dangerous to life or health) describes the atmospheric concentration of any toxic, corrosive, or asphyxiant substance that poses an immediate threat to life or would interfere with a person's ability to escape from a dangerous atmosphere.

The answer is B.

1.3. First responders at the awareness level as identified in 29 CFR 1910.120(q)(6)(i) are individuals who are likely to witness or discover a hazardous substance release and who have been trained to initiate an emergency response sequence by notifying the authorities of the release.

The answer is D.

1.4. 29 CFR 1910.120(e)(3)(i) requires 40 h of off-site training for general site workers.

The answer is D.

1.5. In response to a hazardous substance release, the HAZWOPER Emergency Response Compliance Directive indicates that process operators are authorized to perform only the limited action specifically addressed in the emergency response plan.

The answer is C.

SOLUTION 2

2.1. Container labeling, material safety data sheets, and employee training are all part of the Hazard Communication Standard, but personal exposure monitoring is not.

The answer is B.

2.2. A combustible liquid is defined under the Hazard Communication Standard as one with a flash point between 38°C (100°F) and 93°C (200°F).

The answer is B.

2.3. Manufacturers and importers are responsible for evaluating chemicals to determine if they are hazardous.

The answer is D.

2.4. Employers are required to provide training to employees on hazardous chemicals in their work areas when employees are initially assigned to a work area.

The answer is A.

2.5. The terms "corrosive," "irritant," and "sensitizer" are all used to designate a chemical as a health hazard under the Hazard Communication Standard, but "reactive" is not.

The answer is C.

SOLUTION 3

3.1. E_c = cumulative exposure, ppm
T = time of exposure, h
C = concentration of chemical for time 1, 2 and 3, ppm

$$E_c = \frac{C_1 T_1 + C_2 T_2 + C_3 T_3}{8 \text{ h}}$$

$$E_{c,\text{acetone}} = \frac{\left(\begin{array}{c} (1210 \text{ ppm})(1 \text{ h}) + (719 \text{ ppm})(2 \text{ h}) \\ + (370 \text{ ppm})(5 \text{ h}) \end{array}\right)}{8 \text{ h}}$$

$$= \boxed{562 \text{ ppm} \quad (560 \text{ ppm})}$$

The answer is A.

3.2. $E_{c,\text{toluene}} = \dfrac{(380 \text{ ppm})(3 \text{ h}) + (140 \text{ ppm})(5 \text{ h})}{8 \text{ h}}$

$= 230 \text{ ppm}$

E_m = cumulative exposure to mixture, unitless
L = TWA PEL for acetone and for toluene, ppm

$$E_m = \frac{E_{c,\text{acetone}}}{L_{\text{acetone}}} + \frac{E_{c,\text{toluene}}}{L_{\text{toluene}}}$$

$$= \frac{560 \text{ ppm}}{1000 \text{ ppm}} + \frac{230 \text{ ppm}}{200 \text{ ppm}}$$

$$= \boxed{1.7}$$

The answer is B.

3.3. Because the cumulative exposure to the acetone and toluene mixture is greater than 1.0, the exposure is not acceptable.

The answer is D.

3.4.

$$E_c = \frac{\left(\begin{array}{l}(480 \text{ ppm})(10 \text{ min})\left(\dfrac{1 \text{ h}}{60 \text{ min}}\right) \\ + (380 \text{ ppm})\left(3 \text{ h} - (10 \text{ min})\left(\dfrac{1 \text{ h}}{60 \text{ min}}\right)\right) \\ + (140 \text{ ppm})(5 \text{ h})\end{array}\right)}{8 \text{ h}}$$

$$= \boxed{232 \text{ ppm} \quad (230 \text{ ppm})}$$

The answer is A.

3.5. The cumulative exposure when the toluene concentration peaks at 480 ppm for 10 min during the first 3 hours is 230 ppm. This is greater than the TWA PEL of 200 ppm; therefore, the exposure is not acceptable.

The answer is D.

SOLUTION 4

4.1. Setbacks from mechanical equipment are not included in general requirements for hazardous materials storage.

The answer is B.

4.2. 5 ft is the minimum setback.

The answer is A.

4.3. A Class IV organic peroxide burns as an ordinary combustible and presents a minimum reactivity hazard.

The answer is B.

4.4. A Class I unstable, non-water reactive material is normally stable.

The answer is D.

4.5. Cryogenic fluids are classified as "corrosive/highly toxic," "flammable," or "nonflammable," but not as "reactive/unstable."

The answer is D.

SOLUTION 5

5.1. The PSM standard applies to a flammable liquid or gas in a quantity of 10,000 lbm or more except when the material is used exclusively for workplace consumption as a fuel.

The answer is C.

5.2. Normally unoccupied remote facilities, oil or gas well drilling or servicing operations, and retail facilities are not subject to the PSM standard.

The answer is D.

5.3. The process hazard analysis should be updated and revalidated at least at 5 yr intervals after completing the initial process hazard analysis.

The answer is B.

5.4. Employee refresher training should be provided under the PSM standard as needed, but at least every 3 yr.

The answer is D.

5.5. A compliance audit is prompted under the PSM standard when 3 yr have elapsed since the previous audit.

The answer is C.

SOLUTION 6

6.1. When employees provide their own PPE, the employer must ensure the equipments' adequacy, including proper maintenance and sanitation.

The answer is D.

6.2. To ensure that employees are proficient in the use of PPE, employers must provide training, require the employee to demonstrate proficiency, retrain employees who lose proficiency, and provide written certification that these things have occurred.

The answer is D.

6.3. Of the choices listed, only a positive-pressure self-contained breathing apparatus would be suitable for entering a space that is immediately dangerous to life and health (IDLH).

The answer is C.

6.4. Under conditions where chemical-protective clothing (CPC) is required, CPC with a high "clo" value would be most suitable for lower work rate and lower temperature.

The answer is A.

6.5. USEPA level of protection A defines the highest degree of respiratory, skin, and eye protection.

The answer is A.

6.6. Gender is probably not a factor in evaluating a worker's susceptibility to heat stress while wearing chemical-protective clothing.

The answer is B.

SOLUTION 7

7.1.
$$h = \text{dose-equivalent, rem}$$
$$D = \text{absorbed dose, } 0.012 \text{ rad}$$
$$Q_F = \text{quality factor, unitless} = 20$$

$$h = DQ_F = (0.012 \text{ rad})(20)$$

$$= \boxed{0.24 \text{ rem}}$$

The answer is B.

7.2. $h = (0.32 \text{ rad})(1) = \boxed{0.32 \text{ rem}}$

The answer is D.

7.3. Since the beta radiation dose-equivalent of 0.32 rem is greater than the alpha radiation dose-equivalent of 0.24 rem, the beta radiation presents the greater hazard to the exposed individual.

The answer is D.

7.4.
$$D = \frac{h}{Q_F} = \frac{0.32 \text{ rem}}{20}$$

$$= \boxed{0.016 \text{ rad}}$$

The answer is B.

SOLUTION 8

8.1. The unit used to express radioactivity is the curie, Ci. Roentgen, R, is used for exposure, rad is used for absorbed dose, and rem is used for dose-equivalent.

The answer is C.

8.2. In air, alpha particles can travel a few inches, beta particles can travel up to about 30 m, and gamma radiation can travel several thousand meters.

The answer is A.

8.3. Dose-equivalent is the product of absorbed dose and a unitless conversion factor, called the quality factor, specific to the type of radioactivity.

The answer is B.

8.4. Alpha particles present the greatest health hazard from exposure through inhalation and ingestion.

The answer is B.

8.5. Gamma radiation can penetrate several centimeters of lead shielding and a few meters of concrete. Consequently, there is no protective clothing that will shield individuals from gamma radiation and still allow them to move about.

The answer is D.

8.6. For unprotected individuals, radiation exposure can occur without contamination. Exposure occurs when the body is subjected to radiation emitted from radioactive material regardless of whether the material physically touches the individual. Contamination, however, only occurs when actual radioactive material has become attached to the body or clothing.

The answer is C.

SOLUTION 9

9.1. Microwave, radio wave, and ultraviolet light are all examples of nonionizing radiation. A medical diagnostic x-ray is an example of ionizing radiation.

The answer is A.

9.2. The greatest worldwide contributors to ionizing radiation exposure are natural background sources including cosmic radiation, terrestrial radiation, and naturally occurring radio nuclides. Defense industry and stockpiles, medical applications including x-rays, and nuclear reactor wastes (all sources) are not part of natural background radiation sources.

The answer is C.

9.3. Altitude exerts the most overall influence on cosmic radiation exposure at the earth's surface. Latitude has an overall lesser influence, and weather and dress and outdoor activity are not factors.

The answer is A.

9.4. The three major series of radioactive elements are the actinium series, the thorium series, and the uranium series. Radium and polonium are isotopes occurring in the uranium series.

The answer is B.

9.5. The effective half-life of a nuclide is the lesser of the physical half-life and the biological half-life. Since the given physical half-life of 15 h is less than the given biological half-life of 11 d, the effective half-life is 15 h.

The answer is A.

SOLUTION 10

10.1. 40 CFR 370.2 includes immediate (acute) health hazards, delayed (chronic) health hazards, and sudden releases of pressure as hazard categories, but not corrosives, other than immediate health hazard.

The answer is C.

10.2. The basic requirement for MSDS reporting to the local emergency planning committee as stated in 40 CFR 370.21 is that regulated facilities shall submit an MSDS for each hazardous chemical present at the facility.

The answer is A.

10.3. Tier II chemical inventory forms include specific information listed by chemical such as chemical names and CAS numbers and Tier I forms do not.

The answer is C.

10.4. 40 CFR 370.30 states that any person may request MSDS information for a regulated facility and expect a response from the local emergency planning committee.

The answer is A.

10.5. Failure of a regulated facility to comply with MSDS or inventory reporting requirements may result in monetary civil and administrative penalties as specified in 40 CFR 370.5.

The answer is C.

SOLUTION 11

11.1. "CERCLA hazardous substance," "extremely hazardous substance," and "hazardous chemical" define specific groups of materials under different regulatory citations.

The answer is C.

11.2. The threshold planning quantity (TPQ) is the quantity of a regulated material handled at a facility that, when equaled or exceeded, subjects the facility to regulation.

The answer is B.

11.3. Of the two values listed for threshold planning quantity, the first applies to materials with higher hazard properties and the second applies to materials with lower hazard properties as described under 40 CFR 355.30.

The answer is C.

11.4. As indicated in 40 CFR 355.40, emergency response notification is not required when the release results in exposure to persons solely within the boundaries of the facility.

The answer is B.

11.5. The name and title of the person reporting the release is not included as part of emergency release notification under 40 CFR 355.40.

The answer is C.

11.6. Failure to report a release subject to the regulations may result in monetary civil and criminal penalties or imprisonment as specified in 40 CFR 355.50.

The answer is D.

SOLUTION 12

12.1. Oil, gasoline, and paint are examples of materials that would be involved in an NFPA Class B fire.

The answer is B.

12.2. An NFPA Class II combustible liquid is defined by a flash point greater than or equal to 100°F (38°C) but less than 140°F (60°C).

The answer is B.

12.3. A rapid pressure increase in a confined vessel that results in a sudden release as the vessel ruptures is an explosion.

The answer is D.

12.4. Foam agents are suitable for suppressing NFPA Class A and B fires.

The answer is B.

12.5. A fire involving reactive materials would likely not be appropriately suppressed using water.

The answer is C.

SOLUTION 13

13.1. The most important parameters commonly used to characterize sound are pressure, power, and intensity.

The answer is B.

13.2. The common unit of noise is the decibel (dB).

The answer is A.

13.3. The audible frequency range for humans is between 16 Hz and 20 000 Hz.

The answer is C.

13.4. Solar radiation is typically not considered a direct noise-attenuating factor.

The answer is C.

13.5. Airport noise is generally associated with perceived noise level (PNL), composite noise rating (CNR), and noise exposure forecast (NEF).

The answer is A.

SOLUTION 14

14.1. L_w = sound power level, dB
P_w = sound power = 50 W

$$L_w = 10 \log \left(\frac{P_w}{10^{-12} \text{ W}} \right)$$

$$= 10 \log \left(\frac{50 \text{ W}}{10^{-12} \text{ W}} \right)$$

$$= \boxed{137 \text{ dB} \quad (140 \text{ dB})}$$

The answer is D.

14.2. Assume sound radiates from the source in all directions.

I = sound intensity, W/m^2
A = area of air receiving noise normal to the direction of propagation
  = surface area of a sphere with 100 m radius
  = $4\pi(100 \text{ m})^2$
  = $125\,600 \text{ m}^2$

$$I = \frac{P_w}{A} = \frac{50 \text{ W}}{125\,600 \text{ m}^2}$$

$$= \boxed{3.98 \times 10^{-4} \text{ W/m}^2 \quad (4.0 \times 10^{-4} \text{ W/m}^2)}$$

The answer is A.

14.3. L_I = sound intensity level, dB

$$L_I = 10 \log \left(\frac{I}{10^{-12} \text{ W}} \right)$$

$$= 10 \log \left(\frac{4.0 \times 10^{-4} \dfrac{\text{W}}{\text{m}^2}}{10^{-12} \text{ W}} \right)$$

$$= \boxed{86 \text{ dB}}$$

The answer is A.

14.4. The sound pressure level is

$$L_p \approx L_I = \boxed{86 \text{ dB}}$$

The answer is A.

14.5. At 1000 Hz, 1 dB (sound pressure level) = 1 phon.

S = loudness level, sones
P = loudness level = 86 dB at 1000 Hz = 86 phons

$$S = 2^{(P-40)/10} = 2^{(86-40)/10}$$

$$\boxed{= 24.25 \text{ sones} \quad (24 \text{ sones})}$$

The answer is A.

SOLUTION 15

15.1. ISO 14000 is a set of voluntary environmental standards applied to private-sector business and industry worldwide.

The answer is D.

15.2. The two general categories into which ISO 14000 standards are divided are 1) organization evaluation standards and 2) product and process evaluation standards.

The answer is A.

15.3. Environmental site assessment (ESA) is not included as a major part of the ISO 14000 standards.

The answer is C.

15.4. The three possible outcomes of the ISO 14000 registration process are approval, conditional approval, and disapproval.

The answer is B.

SOLUTION 16

16.1. All chemicals manufactured for sale must appear on the TSCA inventory list.

The answer is B.

16.2. FIFRA requires that only pesticides registered with the USEPA can be distributed, sold, or placed into commerce.

The answer is A.

16.3. General-use pesticides are pesticides that can be used by untrained persons according to label instructions without creating unreasonable adverse effects to the person applying the pesticide or to the environment.

The answer is D.

SOLUTION 17

17.1. Among other topics, the Life Safety Code® covers building services and fire protection equipment.

The answer is A.

17.2. The hazard of contents classification employed in the Life Safety Code® is low, ordinary, and high.

The answer is B.

17.3. Of the choices listed, only staircase dimensional criteria are addressed under means of egress in either the Life Safety Code® or the UBC.

The answer is B.

17.4. The UBC group classification for hazardous occupancy is Group H.

The answer is C.

17.5. The UBC classification for interior finishes with a flame-spread index of less than 25 is Class I.

The answer is B.